# Toxicology in the Middle Ages and Renaissance

# Toxicology in the Middle Ages and Renaissance

Edited by

Philip Wexler

ACADEMIC PRESS
An imprint of Elsevier
elsevier.com

Academic Press is an imprint of Elsevier
125 London Wall, London EC2Y 5AS, United Kingdom
525 B Street, Suite 1800, San Diego, CA 92101-4495, United States
50 Hampshire Street, 5th Floor, Cambridge, MA 02139, United States
The Boulevard, Langford Lane, Kidlington, Oxford OX5 1GB, United Kingdom

**Notices**
Knowledge and best practice in this field are constantly changing. As new research and experience broaden our
understanding, changes in research methods, professional practices, or medical treatment may become
necessary.

Practitioners and researchers must always rely on their own experience and knowledge in evaluating and using
any information, methods, compounds, or experiments described herein. In using such information or methods
they should be mindful of their own safety and the safety of others, including parties for whom they have a
professional responsibility.

To the fullest extent of the law, neither the Publisher nor the authors, contributors, or editors, assume any
liability for any injury and/or damage to persons or property as a matter of products liability, negligence or
otherwise, or from any use or operation of any methods, products, instructions, or ideas contained in the
material herein.

**British Library Cataloguing-in-Publication Data**
A catalogue record for this book is available from the British Library

**Library of Congress Cataloging-in-Publication Data**
A catalog record for this book is available from the Library of Congress

ISBN: 978-0-12-809554-6

For Information on all Academic Press publications
visit our website at https://www.elsevier.com/books-and-journals

Working together
to grow libraries in
developing countries

www.elsevier.com • www.bookaid.org

*Publisher:* Mica Haley
*Acquisition Editor:* Erin Hill-Parks, Rob Sykes
*Editorial Project Manager:* Tracy I. Tufaga
*Senior Production Project Manager:* Priya Kumaraguruparan
*Designer:* Matthew Limbert

Cover image credit: An alchemist in his laboratory. Oil painting by a follower of David Teniers the younger.
Wellcome Library, London

Typeset by MPS Limited, Chennai, India

"For Nancy, Yetty, Will, Jake, Lola, and Gigi, with Love"

# CONTENTS

# LIST OF CONTRIBUTORS

**Mohammad Abdollahi**
Faculty of Pharmacy and Pharmaceutical Sciences Research Center, Tehran University of Medical Sciences, Tehran, Iran

**Mozhgan M. Ardestani**
School of Traditional Medicine, Tehran University of Medical Sciences, Tehran, Iran

**Sheila Barker**
The Medici Archive Project, Florence, Italy

**Steven Bednarski**
St. Jerome's University in the University of Waterloo, Waterloo, ON, Canada

**Cathy Cobb**
University of South Carolina, Aiken, SC, United States

**Maria do Sameiro Barroso**
Portuguese Medical Association, Lisbon, Portugal

**Mike Dash**
University of Cambridge, Cambridge, United Kingdom

**Christopher J. Duffin**
The Natural History Museum, London, United Kingdom

**Benedetta F. Duramy**
Golden Gate University School of Law, San Francisco, CA, USA

**Mohammad M. Esfahani**
School of Traditional Medicine, Tehran University of Medical Sciences, Tehran, Iran

**Gabriele Ferrario**
Cambridge University Library, Cambridge, United Kingdom

**Urs Leo Gantenbein**
University of Zurich, Zurich, Switzerland

**Frederick Gibbs**
University of New Mexico, Albuquerque, United States

**Omar Habbal**
Sultan Qaboos University, Muscat, Sultanate of Oman

**Georgiana D. Hedesan**
University of Oxford, Oxford, United Kingdom

**Sverre Langård**
Oslo University Hospital, Oslo, Norway

**Chris Lavers**
University of Nottingham, Nottingham, United Kingdom

**Caley McCarthy**
McGill University, Montreal, QC, Canada

**Roja Rahimi**
School of Traditional Medicine, Tehran University of Medical
Sciences, Tehran, Iran

**Alain Touwaide**
Institute for the Preservation of Medical Traditions, Washington, DC,
United States

**Kathleen Walker-Meikle**
University College London, London, United Kingdom

Following on the heels of this series' first two volumes on *Toxicology in Antiquity*, the current installment covers the approximate dates of 400−1600. Although lacking in consensus, the Middle Ages is often thought to begin about the time of the fall of Rome in 476. Stirrings of the Renaissance were apparent in the mid-1300s and it began to flourish with the fall of Constantinople in 1453.

This volume presents tales of famous alleged poisoners such as the Borgias, the Medicis, Giulia Tofana, and Catherine Monvoisin, although not all their reputations for poisoning are merited. We learn, for example, that while the Borgias were no saints, they were hardly the poisoners that history has made them out to be. The less celebrated Margarida de Portu makes an appearance as the defendant in a late 14th century poisoning courtroom trial. A chapter on animal venoms illustrates the continued interest in this topic since ancient times. Readers will be interested to know that in the Middle Ages, the saliva of a rabid dog was considered to be a poison every bit as potent as snake venom. Hand in hand with actual poisoning episodes of the period was the ever elusive search for antidotes. Without much to substantiate efficacy claims, organic substances such as animal horns, internal animal stones (such as bezoars), and fossil shark teeth were touted both to prevent and cure poisoning. Several chapters are devoted to these topics.

Despite the unabated public fascination, since ancient times, with poisons, poisoners, and poisoning victims, true scientific milestones were also reached during the period covered by this volume. Portions of the pseudo-science of alchemy evolved into the true science of chemistry. A chapter on Jan Baptist van Helmont discusses the influence of his alchemical theories on toxicology. Intellectuals of the time made scholarly contributions to an understanding of toxicology, some with more credible scientific underpinnings than others. The great Spanish born Jewish philosopher and physician, Moses Maimonides, wrote a treatise on poisons. Meanwhile, toxicology was a featured science during the Islamic Golden Age, particularly in Persia. Scientists such as

Avicenna and Rāzī, and their successors, did much to stimulate its growth in the Middle East, with subsequent spread throughout Europe. The Italian, Pietro d'Abano's *Liber de Venenis* represents a recapturing of toxicology's Greek origins. Georgius Agricola, from Saxony, the "father of minerology," contributed to the systematic study of mining and the diseases of miners. And, of course, Paracelsus encapsulated so much about the toxic action of substances in a few sentences, which tend to be abbreviated to "The dose makes the poison." While a bit of a simplification measured against current scientific standards, this remains a powerfully true adage today. Urs Leo Gantenbein who leads the ambitious Paracelsus Project at the University of Zurich outlines this Renaissance figure's significance to toxicology.

During the Middle Ages and Renaissance, science made significant strides in learning how toxic agents interact with the body. The authors of the chapters herein are to be congratulated on bringing to life a host of characters of toxicological significance, both nefarious practitioners of poisoning, and brilliant scientists and scholars of the time who contributed to a budding understanding of toxicology.

**Philip Wexler**

# ACKNOWLEDGMENTS

With thanks to Tracy Tufaga, Erin Hill-Parks and Priya Kumaraguruparan of Elsevier.

# ABOUT THE EDITOR

Philip Wexler, of the National Library of Medicine's Toxicology and Environmental Health Information Program, is a member of the Society of Toxicology and a recipient of its Public Communications Award (2010). He serves as federal liaison to the Toxicology Education Foundation. He has lectured and published widely in toxicology information and history, and edited numerous publications. In addition to serving as editor-in-chief for this series, he has served as editor-in-chief for the *Encyclopedia of Toxicology*, 3rd ed. (Elsevier, 2014), *Information Resources in Toxicology*, 4th ed (Elsevier, 2009. new edition in preparation), *Chemicals, Environment, Health: A Global Management Perspective* (Taylor and Francis, 2011) and the journal, *Global Security: Health, Science, and Policy* (Taylor and Francis/ Routledge).

# CHAPTER *1*

# Poison and Its Dose: Paracelsus on Toxicology

**Urs Leo Gantenbein**
University of Zurich, Zurich, Switzerland

## 1.1 THE FOUR PILLARS OF MEDICINE

The adage, "The dose makes the poison," is perhaps the most famous quote in the history of toxicology. It was coined by the Swiss physician, natural philosopher, and radical church reformer Theophrastus Bombast of Hohenheim, called Paracelsus (1493/94–1541). Being an enigmatic and independent Renaissance thinker, he was long misunderstood and a subject of controversy, in part even to this day. Paracelsus intended nothing less than to completely reform medicine which in his day was founded on mere book learning and the rigid medical doctrines of the ancients. As brilliant as his mind was, so was his character difficult and unrelenting, especially when confronted with dissent, which often was the case. As a result of orthodox physicians' opposition, his life was one of the endless wanderings throughout Europe with local authorities reluctant to provide him license to stay at a place longer than a few months. Paracelsus reached the climax of his career in Basle, Switzerland, where in 1527 he was appointed town physician and professor at the university. He announced plans to revise the ancient doctrines, to advocate for the importance of practical experience in medical matters, and to establish a new framework of medicine and surgery in books he would write himself. In just a few months, he presented an extensive series of lectures covering pharmacology, medicine, and surgery. After an ideological clash with his students and a legal dispute with his superiors, he was forced to flee from the town, a blow which he never recovered from.

Following this fracas, he composed many other works among which his two famous theoretical writings were the *Paragranum* (1529) and the *Opus Paramirum* (1531) [1]. The former contains the concept of the Four Pillars of Medicine as philosophy, astronomy, alchemy, and virtue (ETW 8-14).[2]. Although natural philosophy, as understood in this scheme,

Toxicology in the Middle Ages and Renaissance. DOI: http://dx.doi.org/10.1016/B978-0-12-809554-6.00001-9

describes the principles and substances found in nature applied to medical therapy, astronomy elucidates the secret bonds between heaven and earth, leaning on the neo-Platonic conception that the stars are celestial ideas which influence and determine everyday occurrences. Through his acquaintance with mining and metallurgy, Paracelsus became knowledgeable in Medieval alchemy. From this, he appropriated the principle of "separatio puri ab impuro" (S 4:115, 132) [3], the separation of the pure essence from an otherwise toxic substance, and applied it to the preparation of effective remedies [4]. In this sense, Paracelsus writes in the *Paragranum*:

> Who is there who would deny that in all good things poison also resides? Everyone must acknowledge this. This being the case, the question I ask is: must one then not separate the poison from what is good, taking the good and leaving what is bad? Of course one must. (ETW 247)

Extending additionally the alchemical motto of "probing in the fire" (ETW 311) to clinical experience, alchemy became the very basis of Paracelsus' medical revolution. Other than the common doctors who used mixtures of mostly unprocessed raw substances, he enriched the remedy treasure with alchemically prepared and thus detoxified mineral drugs. Finally, the fourth pillar, virtue, represents medical ethics, a subject in which Paracelsus was far ahead of his time [5]. Stressing empathy he said, "the greatest foundation of medicine is love" (S 7:369) and "where is no love there is no art" (S 8:263). He was opposed to forsaking the sick in hopeless cases but rather argued for compassion because "mercifulness is the teacher of the physicians" (S 8:264).

The second one of the above-mentioned theoretical works, the *Opus Paramirum*, contains among many other things the medical application of the three basic philosophical principles of Paracelsian alchemy—mercury, sulfur, and salt. First introduced by Paracelsus in the Basle lectures and not to be confused with the corresponding chemical substances, this trinity comprises the fundamental constituents of all organic and inorganic matter [6]. When a substance is observed in the alchemical furnace, mercury stands for the liquid and volatile or part, sulfur represents the oily and combustible part. Salt, on the other hand, is untouched by fire and remains as a fixed principle in the laboratory vessel. Using wood as an example, Paracelsus states, "Let it burn: that which burns is *sulphur*. What smokes is *mercurius*. What turns to ash [is] *sal*" (ETW 319). Later, in the *Opus Paramirum*, Paracelsus reinterpreted the three principles as physiological processes

*Figure 1.1 Portrait of Paracelsus—U.S. National Library of Medicine, Bethesda, MD, United States.*

in a bodily alchemy, seeing the functioning of the body in terms of sublimation, distillation, circulation, heating, and cooling (S 11:188). Using this theory, Paracelsus explained disease as a malfunction of the three principles. Thereby, he had nothing less in mind but to replace the ancient theory of the four humors—blood, phlegm, yellow, and black bile. With all these new ideas, Paracelsus was shaking the very foundations of contemporary medicine. It is therefore not surprising that he was confronted with massive opposition. As a result of general disrepute, from the enormous collection of manuscripts written by Paracelsus, which will fill up to 30 volumes when completely edited [7], only a minor fraction was printed during his lifetime (Fig. 1.1).

## 1.2 POISON AND THE ALCHEMIST IN THE STOMACH

Another important work of Paracelsus relates to his various concepts of poison, namely, the *Volumen medicinae Paramirum*, which is not identical with the later *Opus Paramirum*. Sketchy and rather inconsistent, it comprises probably an early work written before the Basle period.

It contains the etiological theory of the *Five Entia* as the causes of disease, with which Paracelsus has gained a certain fame [8]. Roughly summarizing, the *ens astrale* signifies the pathogenic influences of the stars, the *ens venale* stands for the poison contained in the food and the air, and the action of the *ens naturale* corresponds to the malfunctioning of the body itself. Resembling a force like a spirit "born out from our thoughts without matter in the living body" (S 1:216), the *ens spirituale* could represent psychosomatic disorders and the influence of witchcraft, both based on imagination. Finally, in *ens deale*, disease is understood as a punishment from God. Andrew Weeks argues that Paracelsus could have derived the five entia from the five different suspected explanations for the ravaging plague epidemics of his times. Indeed, the contemporary plague tracts discussed as possible causes, astronomic events, miasmatic theories with poisonous vapors in the air, humors and natural catastrophes, magic and imagination, and divine punishment [9].

Even the *ens astrale* acts by poison, namely, the "vapors of the planets harm us" (S 1:185), and the poisoned stars "stain the air with their poison, so where the poison arrives, there will be diseases springing up according to the stars' qualities" (S 1:184). In asserting the importance of poisons in the development of illness, he stated that the stars are even more powerful and outwit the earthly poisons:

> No physician should be astonished that there are not so many poisons on earth, but be aware that there are even more are in the stars. And every physician should be aware that no illness comes without poison, because in poison lies the very beginning of every illness. (S 1:185)

In describing the *ens venale* or *ens veneni*, as he also called it, Paracelsus declares that the body is not naturally contaminated with toxins, but can, in fact, be poisoned by taking food. This is so because everything contains an essence and a poison. "Essence is what preserves the human being, poison is what brings him sickness" (S 1:195). So when eating, human beings are automatically subject to poison:

> A human being must eat and drink, because the body which houses his life requires it and cannot do without it. Thus the human being is forced to take his own poison, and illness and death, by eating and drinking. (S 1:191)

However, nature knows how to manage this predicament. Drawing a parallel to the alchemical reactions, he observed in his laboratory, Paracelsus imagined that in the human body, there must be a similar assimilating and transforming principle. This "inner alchemist" or "archaeus" (S 11:188), as he later called it, skillfully separates the poison from the food and eliminates the waste from the body:

> Now remember ... while poison may and does harm us, we have got an alchemist in us, installed and given us by the Creator, who shall separate the poison from the good so that we may not receive any disadvantage. (S 1:193)

Human beings and the animals have different inner alchemists with their own capacities to neutralize various poisons. For example, the alchemist of the peacock, which resembles no other animal, is subtle and allows this bird to eat snakes, scorpions, and newts without harm. The alchemist of the ostrich separates iron and the one of the salamander even transmutes fire, but the keenest inner alchemist belongs to the pig which may even ingest human and other animal excrements and extract nourishment from it (S 1:192). Depending on the nature of the poison in question, it is driven out through various openings of the body—the pores of the skin, the nose, the ears, the eyes, the mouth, and via urine and excrement (S 1:199). The alchemist resides in a specific location of the body, namely, in the stomach which naturally resembles a retort which one might find in an alchemist's laboratory. This is where foods are cooked and disentangled. The alchemist transforms the good parts into an essence, and this tincture permeates or "tinges" the whole body in order to produce blood and flesh. When for some reason the alchemist fails in his task, then the precarious food remains in the stomach and is subject to putrefaction, eventually leading to disease:

> He [the inner alchemist] separates the bad part from the good, he transmutes the good into a tincture, he tinges the body that it may live ... that it may become blood and flesh. This alchemist resides in the stomach which is the instrument in which he cooks and works. (S 1:194)
> When the alchemist fails, so that the poison is not be properly separated from the good according to the rules, there is a joint putrefaction of the poison and the good, followed by digestion. This is what results in human disease, because all illnesses obtained by human beings from ens veneni arise from a rotted digestion ... when digestion has failed and the alchemist is not perfect in his instrument, from which corruption follows, and this is the mother of all diseases ... because corruption empoisons the body. (S 1:195)

These quotations amply demonstrate that Paracelsus was eager to find an explanation for the apparently assimilating and excreting processes of the human body. He found the answer from the realm he was accustomed to, i.e., the theory and practice of alchemy with its theories of refining crude materials. Although Paracelsus was not the first to apply alchemy to pharmacology and medicine, he was the first to significantly expand this theory and promote it to a broader audience. In this sense, he anticipated chemical therapeutics, and his efforts to explain vital processes in chemical terms could be considered a precursor of present-day biochemistry.

## 1.3 NOXIOUS MINERAL VAPORS AND THE MINER'S DISEASE

A great many alchemists lived in the vicinity of mining sites [10] and Paracelsus himself had a great interest in mining. The would-be alchemist would do well to travel to mountains and their mines, because "where the minerals lie there are the artists [of alchemy]" (S 11:144). On the occasion of his visits to mines in the Tyrolean and Carinthian alps, Paracelsus had observed that the miners and smelters were afflicted with peculiar diseases. The ores being brought to the surface contained significant portions of toxic substances such as mercury, arsenic, cadmium, antimony, and other heavy metals which were set free when melted or otherwise processed. Due to the absence of significant protective measures, Paracelsus must have been confronted with a great variety of poisonings common in miners. Another kind of illness frequent in unprotected mining and chiefly affecting the lungs is silicosis, which is caused by the constant inhalation of silica dust, or even worse, silico-tuberculosis, in which tuberculosis develops and attacks the already weakened lungs. Paracelsus had this array of diseases in mind when he wrote his lengthy treatise *Von der bergsucht und anderen bergkrankheiten* (on the mine-affliction and other miners' sicknesses) circa 1533 [11]. With its detailed descriptions of poisonings and the corresponding disease patterns, it comes very close to a modern textbook of toxicology. According to Paracelsus, the miners become *bergsüchtig* (mine-afflicted) when they develop lung disease, cachexia, and stomach ulcer. He insists that the ancient writers were unaware of the ailment and relates the various occupational situations where it may arise:

> In order that you know what the mountain-disease is, it is the following: when the ore workers, smelters, miners, and whomever else are kin to the mines, be it in the washing plant, in silver or gold ore, salt ore, alum and sulfur ore, or in vitriol boiling, in lead, copper, tin or quicksilver ore, those who dig in such ores, [when they] fall into lung disease, into emaciation of the body, into stomach ulcer: those are called mine-afflicted. (S 9:463–464)

Paracelsus imagines that the disease is induced by a nebular vapor emanating from the minerals and ores. Not only does this evaporation in the mines sicken workers, but so does the smoke that rises from melting ore. Realizing the nature of chronic intoxications, Paracelsus states that a poisoning of this kind is not acute, but rather a slow process "which pushes people into a long illness" (S 9:478). He gives an illustration with white arsenic (arsenic trioxide $As_2O_3$) which was often found in precipitated smelter smoke:

> As an example: If white arsenic is ingested there is a quick and sudden death, however, when the corpus [substance] is not taken itself but its spiritus [vapor], then an hour is lengthened to a whole year, that is, what the corpus brings about in ten hours is done by the spiritus in ten years. (S 9:478)

He further notes that all the arsenical minerals such as reddish realgar (arsenic tetrasulfide $As_4S_4$) and yellow orpiment (arsenic trisulfide $As_2S_3$) show similar symptoms of poisoning:

> The ingestion of realgar produces a desiccated lung ... wheezing with discoloration of the face, as well as cracks and fissures in the liver, accompanied by an unnatural thirst, and gnaws and grinds the folds of the stomach ... a difficult hard digestion. After that, there is a great deal of heat, palpitating and trembling in the pit of the stomach, followed by convulsions of all the limbs, skin darkening and concomitant headaches. (S 9:478–479)

This remarkably accurate description of chronic arsenic intoxication covers many of the clinical signs known today including thirst, gastrointestinal problems, neurological effects, and the typical discoloration of the skin. There is even a hint of possible liver and stomach cancer. In a similar way, Paracelsus relates the clinical pictures of the intoxications caused by two other mineral classes, represented by *antimonium* and *alkali*. The former comprises insoluble minerals like antimonite, marcasite, steatite, and the like, while the latter includes soluble minerals like vitriols, alums, and other salts. However, the afflicted miners are not left without hope, and here Paracelsus again resorts to the wonders of alchemy. In the same way as with poisonous plants or food, the toxic part of the minerals can be separated, leaving an almost mysterious remedy, which Paracelsus calls *arcanum*. The same substance which has caused jaundice may also heal the jaundice, when purified and turned into an *arcanum* (S 9:481). Related to arsenic intoxication, Paracelsus has argued in a draft of the *Paragranum* in the same vein:

> When you know the arsenicum [white arsenic] and its nature, then you also know how to detect the arsenicum in the body. Now you truly know all about the kind, quality, essence, origin, and nature of the excretions. Having this, you are shown the remedy, because arsenicus heals arsencium, anthrax [coal, ember] anthracem, namely poison heals poison. (S 8:120; see also [12])

The tract on the miner's disease closes with a lengthy theoretical discussion of metallic mercury. Regarding the symptoms of intoxication, Paracelsus is more precise in his tract on syphilis *Von der französischen Krankheit drei Bücher Para* (On the French Disease Three Books Para, 1529). There, he criticizes the drastic mercury therapy applied to syphilitics during that period. The infected persons were smeared with a mercurial ointment or set in sweatboxes where they had to inhale mercury vapors mixed with other ingredients. Paracelsus reports on the symptoms which ensued. The poor patients had to endure heavy mucus oozing from the mouth and appearing in stool. The gingiva and the uvula began to "rot," causing bad breath and the loss of the teeth. The lung was burnt and other organs like the stomach, the kidneys, and the liver were ruined (S 7:79–81).

The above extracts from Paracelsus' writings constitute only a small portion of his comments on intoxications. To sum up, it can be said that he had a remarkable knowledge of toxic minerals and metals. In 1990, the human remains of Paracelsus were subjected to a thorough anthropological analysis [13]. His tomb is situated in the St. Sebastian Cemetery in Salzburg. It is an ironic twist of fate that the investigators found in the bones an unusually high level of mercury, ten to a hundred times more than in normal specimens. This chronic mercury intoxication might have led to his untimely death, which is not surprising since Paracelsus is reported to have almost constantly experimenting with alchemy, and mercury compounds were among his favorites. Living to the age of only 48, his jawbones showed dramatic tooth loss in the last few months of his life. That could mean that Paracelsus, as he became increasingly frail and near death, may have tried to cure his illness with mercury preparations.

## 1.4 THE DOSE MAKES THE POISON

Toward the end of his life, Paracelsus felt the urge to summarize his teachings and defend them before his adversaries. One of his endeavors resulted in the treatise *Die Verantwortung vber etliche Vnglimpfungen seiner Mißgönner* (The defense against several vilifications by his enviers, 1537/38), shortly called the *Seven Defenses*. Alluding to his alchemical remedies, one of the accusations concerned his new recipes:

*[The ignorant doctors] say that the recipes which I prescribe are poison, corrosive, and an extraction of all malignity and virulence of nature. (S 11:136).*

Paracelsus retorts that the common doctors and pharmacists are no different, using poisons like mercury, snake venom, or drastic purgatives, with the important difference that they do not know the art of detoxifying them by alchemy. He reasserts that in every poison lies a great *arcanum*, wherefore toxic substances should not be rejected. "He who despises poison, does not know what is contained in poison" (S 11:137). Then follows the statement which secured Paracelsus a place as the formulator of one of the world's most famous quotes, although rarely understood in its proper context, and practically never cited accurately:

> *When you want to correctly evaluate a poison, what is there that is not poison? All things are poison and nothing is without poison; only the dose determines that something is not a poison.*
>
> *Wenn jhr jedes Gifft recht wolt außlegen/ Was ist das nit Gifft ist? alle ding sind Gifft/ vnd nichts ohn Gifft/ allein die Dosis macht/ dz ein ding kein Gifft ist. (H 2:170; S 11:138)*

# REFERENCES

[1] Theophrastus Bombastus von Hohenheim 1493–1541 (2008). *Essential Theoretical Writings.* Edited and translated with a Commentary by Andrew Weeks. Leiden, Boston: Brill, abbreviated as ETW.

[2] For a description of the Four Pillars of Medicine see ETW 8-14, and also Weeks, A.: Paracelsus: speculative theory and the crisis of early reformation, Albany, 1997, New York State University Press, here 146–157.

[3] Concerning the complete medical and natural philosophic works of Paracelsus the following two editions are generally used: Theophrastus Paracelsus (1589–91). *Bücher und Schrifften,* 10 vols. ed. Johannes Huser. Basle: Conrad Waldkirch. Theophrast von Hohenheim, called Paracelsus (1922–33). *Sämtliche Werke. 1. Abteilung: Medizinische, naturwissenschaftliche und philosophische Schriften,* 14 vols. ed. Karl Sudhoff. Munich and Berlin: Oldenbourg. The two editions are abbreviated H and S respectively together with volume and page numbers. When not taken from ETW, the translations from German to English are done by myself.

[4] Gantenbein, U. L. (2011). Paracelsus und die Quellen seiner medizinischen Alchemie. *Religion und Gesundheit im 16. Jahrhundert,* ed. Albrecht Classen (Theophrastus Paracelsus Studien, 3). Berlin and New York: De Gruyter, 113–164.

[5] Gantenbein, U. L. (1998). Medicus ex Deo: Die ärztliche Ethik des Paracelsus im Licht antiker und mittelalterlicher Traditionen. *Nova Acta Paracelsica N. F. 12,* 220–262.

[6] For a description of the *Opus Paramirum* see ETW 14-23. Concerning the Three Principles see Pagel, W. (1982). *Paracelsus. An Introduction to Philosophical Medicine in the Era of the Renaissance* (2nd ed.). Basel etc.: Karger, here 100–104; Gantenbein, U. L. (1997). Separatio puri ab impuro: Die Alchemie des Paracelsus. *Nova Acta Paracelsica N. F. 11,* 3–59, here 26–39.

[7] Presently a whole fourth of the collected works, all of them theological writings, are not yet printed. The edition of the remaining works will be carried out by the New Paracelsus Edition, see www.paracelsus.uzh.ch.

[8] On the *Five Entia* see Weeks, 1997, 60–75.

[9] See Weeks, *Speculative Theory,* 68–73.

[10] Gantenbein, U. L. (2000). Die Beziehungen zwischen Alchemie und Hüttenwesen im frühen 16. Jahrhundert, insbesondere bei Paracelsus und Georgius Agricola. *Mitteilungen der Fachgruppe Geschichte der Chemie der Gesellschaft Deutscher Chemiker 15,* 11–31.

[11] An English translation of the treatise on the miner's sickness is provided by Sigerist, H. E. (1941). *Four Treatises of Theophrastus von Hohenheim Called Paracelsus.* Baltimore: The John Hopkins Press, pp. 43–126. This translation is not used here.

[12] See ETW 102-103 for a discussion of the somewhat unclear term *anthrax.*

[13] Dopsch, H., & Kramml, P. F. (1994). *Paracelsus und Salzburg.* Salzburg: Gesellschaft für Salzburger Landeskunde. Harrer, G. Zur Todeskrankheit des Paracelsus, 61–67. Kritscher, H., Szilvássy, J., & Vycudilik, W. Die Gebeine des Arztes Theophrastus Bombastus von Hohenheim, genannt Paracelsus. 69–96. Reiter, Ch., Das Skelett des Paracelsus aus gerichtsmedizinischer Sicht, 97–116.

# CHAPTER 2

# The Golden Age of Medieval Islamic Toxicology

Mozhgan M. Ardestani[1], Roja Rahimi[1], Mohammad M. Esfahani[1], Omar Habbal[2] and Mohammad Abdollahi[3]

[1]School of Traditional Medicine, Tehran University of Medical Sciences, Tehran, Iran [2]Sultan Qaboos University, Muscat, Sultanate of Oman [3]Faculty of Pharmacy and Pharmaceutical Sciences Research Center, Tehran University of Medical Sciences, Tehran, Iran

## 2.1 INTRODUCTION

Islamic countries in the Middle Ages made early and important contributions to toxicology. This was mainly due to the need of kings and conquerors to prepare poisons and antidotes for fighting enemies and competitors (Mohagheghzadeh et al., 2006). Sudden death was not an uncommon occurrence in royal courts and was often attributed to poisoning (Tschanz, 2013). Many of the Shiite Imams were killed by poisons. One example is Imam Hassan Ibn Ali Ibn Abi Talib (624−70 CE), who died from repeated and chronic poisoning with arsenic and consequent cirrhosis and gastrointestinal bleeding (Nafisi, 1974). The fear of poisoning convinced Umayyed caliphs (661−750 AD) of the need to study poisons, their identity and methods of detection, as well as the prevention and treatment of poisoning. Animal bites, including the bites of dogs, snakes, scorpions and spiders as well as the poisonous properties of various minerals and plants, such as aconite (*Aconitum napellus*), mandrake (*Mandragora officinarum*), and black hellebore (*Helleborus niger*) were of great interest (Tschanz, 2013).

The antiquity of toxicology in Persian and Arabic countries is apparent in searching for the roots of two words: "toxin" and "bezoar." "Toxin" was derived from the Persian word "taxsa" meaning "poisoned arrow" used in ancient Persia (Davari, 2013). The first known use of the word "toxin" was in 1886. The standard etymology suggests it is derived in 1660 from the French "*toxique*" and from Late Latin *toxicus* meaning "poisoned" and from Latin *toxicum* "poison," from ancient Greek *toxikon* (pharmakon) "poison for use on arrows" and from *toxikon*,

Toxicology in the Middle Ages and Renaissance. DOI: http://dx.doi.org/10.1016/B978-0-12-809554-6.00002-0

neuter of *toxikos* "pertaining to arrows or archery," and from *toxon* "bow," probably from a Scythian word that was borrowed by Latin as *taxus* the "yew" plant. Watkins, however, suggested a possible Persian source *taxša-* "bow," from Proto-Indo-European root *tekʷ-os* "to run/ flow," it was used as a noun from 1890 (Klein, 1971; Weekley, 1921). The word "bezoar" was derived from the Persian "pād-zahr" which means a stone with antidotal properties (Mohagheghzadeh et al., 2006). The subject of "antidote" has a particular place in traditional Islamic medicine and its mention in literature confirms this issue. **Avicenna** (*Abu Ali al-Husayn ibn Abdullah ibn Al-Hasan ibn Ali ibn Sina, 983−1035 CE*), **Imād al-Din Shirāzi** (*Imad al-Din Mahmud ibn Mas'ud Shirazi, 1515−92 AD*), and **Burhan al-Din Nafis ibn Ewaz Kermāni** (*15th century physician, wrote a valuable commentary on Najib al-Din Samarqandi's Ketāb al-asbāb wa al-alāmāt and enriched its section on ophthalmology*) are among the Islamic scholars who have specifically addressed this issue.

## 2.2 PROMINENT TOXICOLOGISTS IN MEDIEVAL ISLAMIC ERA

Numerous Islamic authors discussed poisons and their antidotes. Among them was **Ibn Uthal** (seventh century), the Christian physician to the first Umayyad caliph, Mu'awiyah. He was a noted alchemist and skilled toxicologist who conducted a systematic study of antidotes and poisons. Another Christian physician−toxicologist, **Abu al-Hakam al-Dimashqi**, served the second Umayyad caliph, Yazid (Tschanz, 2013; Shahîd, 2010). Other prominent Islamic toxicologists are highlighted below.

### 2.2.1 Jābir (Jaber)-ibn-Hayyān (721−815 AD)
Abu Mūsā Jābir ibn Hayyān, also known as Geber to Europeans, has made valuable contributions to the field of toxicology; although many of those activities were placed in the shadow of his prominent position in the field of chemistry (Fig. 2.1). Jāber claimed that he wrote about 500 treatises though only a few are available now. A very large body of Arabic writings, many of them are highly tantalizing, were authored by Jâbir ibn Hayyân. It was this enormous scope of the Jâbirian corpus and its size that constituted largely, though not exclusively, the grounds for Paul Kraus to conclude that these texts were not the work of a single author, and that they were written no earlier than the 9th century, and took some 100 years to complete. This view, expressed by

*Figure 2.1 Portrait of Jābir-ibn-Hayyān "Geber," Codici Ashburnhamiani 1166, Biblioteca Medicea Laurenziana, Florence, Italy.* Adapted from: https://en.wikipedia.org/wiki/Jabir_ibn_Hayyan.

a scholar who still remains the greatest modern authority on Jâbir, has generally been accepted by the bulk of contemporary historians of science in the West. As for the large number of actual texts making up the Jâbirian corpus, it was a monumental contribution of Paul Kraus that he carried out an exhaustive census of these writings, and ordered them chronologically (Kraus, 1942–1943).

One of the books attributed to Jâbir is "al-Somum" which is the most detailed and richest book in the field of toxicology among Arabic and Greek writings. Other than definitions, types, dose of administration, and constituents of poisons, this book contain medical and therapeutic theories. He criticized the pharmacological and toxicological viewpoints of Hippocrates and Galen. For example, in a part of this book, Jābir was critical of Galen for the administration of cashew (*Anacardium occidentale*) shell oil for some diseases. The manner in which Jābir critiqued some previously accepted wisdom demonstrates his proficiency in the field of toxicology (Sezgin, 2001).

### 2.2.2 Ibn Māsawyah (Yuhanna ibn Masawyah, Abu Zakariya, 777–857 AD)

Yūḥannā Ibn Māsawyah was a pharmacist in Gundishāpur in Western Iran (the intellectual center of the Sassanid Empire). One of his books is "al-Somum," a book on toxicology, known to Europeans as *Mesué* or *filius Mesué*.

### 2.2.3 Al-Kindi (805–73 AD)

Abu Yūsuf Ya'qūb ibn Is āq aṣ-Ṣabbā al-Kindī, a great Arabic philosopher and mathematician at the time of the Abbasid Caliphs, has written several treatises on toxicology, known to Europeans as *Al Kindus*.

### 2.2.4 Qusṭā ibn Lūqā (820–912 AD)

Qusṭā ibn Lūqā al-Ba'labakkī was an Arabic mathematician who wrote "Fi Dafe Zarar al-Somum," i.e., "how to detoxify poisons" (Sezgin, 2001).

### 2.2.5 Ibn Waḥshīyah al-Nabti (9–10th century AD)

Abu Bakr, Ahmad Ibn Ali Ibn Waḥshīyah, wrote a book specifically about toxicology, "Al-Somum Wa al-Taryaghat" (Fig. 2.2). This book is a translation of the Nabataean book "Bartugha," but it is not just a simple translation; he also introduced his own experiences and knowledge about toxins. In the introduction, he mentions other books that had been written about toxicology in India, Persia, Greece, and elsewhere. The author expertly advised that common people or those who don't possess enough knowledge should not utilize this book because of the information about poisons that exist in the book and that it might be misused by people. This is an indication of their close attention to ethical issues in medicine. In a part of the mentioned book, he classifies poisons based on their mode of administration. For example, their toxic effects appear after either touching, inhaling, or ingesting. In its entirety, this book could be divided into two major parts: (1) Toxins and their types; (2) a variety of ways to deal with poisoning (Sezgin, 2001, Mousavi Bojnourdi, 1988). The book was translated into English with a useful introduction and indexes by Martin Levey as "Medieval Arabic Toxicology, the Book on Poisons of Ibn Wahshiyya and Its Relation to Early Indian and Greek Texts." He described hashish (hemp and marijuana) in one part of his book:

*Figure 2.2 "Al-Somum va al-Tary aghat" written by Ibn Wahshia in 909 AD.* Adapted from: Library, Museum and Document Center of Iran Parliament: http://ical.ir/.

*"... If it reaches the nose, a violent tickle occurs in the nose of this man, then in his face. His face and eyes are affected by an extreme and intense burning; he does not see anything and cannot say what he wishes. He swoons, then recovers, swoons [again], and recovers [again]. He goes on in this way until he dies. A violent anxiety and fainting goes on until he succumbs, after a day, a day and a half, or more. If it is protracted, it may take two days. For these aromatics, there is no remedy. But if God wills to save him, he may be spared from death by the continuance of vomiting or by another natural reaction..."* (Levey, 1966).

## 2.2.6 Al-Rāzī (865–925 AD)

**Abū Bakr Muhammad ibn Yahya ibn Zakariyyā al-Rāzī** known as Rāzī in Western countries (Figs. 2.3 and 2.4) was one of the great philosophers and physicians of Persia. His book *al-Hawi*, completed after his death, is a compilation of his working notebooks, which included observations gathered from other books as well as original notes on diseases and therapies, based on his own clinical experience. In volume XIX of "al-Hawi," the most detailed book of Razi, bites and poisons are specifically mentioned (Figs. 2.5–2.7). In that volume, patients with animal bites, including dog, monkey, and horse, and the related symptoms and treatments are discussed. Moreover, stings of flies, bees, and other insects and related complications and treatments have been highlighted. Razi provides an interesting discussion in this volume about rabid dogs and symptoms of rabies (Rāzī, 2008a,b). When Razi was asked to find an appropriate place for building a hospital

*Figure 2.3 Imaginary portrait of Rāzī.* Adapted from: http://www.learn-persian.com/english/Razi_Zakariya.php.

*Figure 2.4 The statue of Razi in United Nations Office in Vienna is part of the "Scholars Pavilion" donated by Iran.* Adapted from: https://en.wikipedia.org/wiki/Muhammad_ibn_Zakariya_al-Razi.

*Figure 2.5 The final page of the copy of the Al-Hawi by Rāzī, with the colophon in which the unnamed scribe gives the date he completed the copy as Friday, the 19th of Dhu al-Qa'dah in the year 487 (=November 30, 1094). The National Library of Medicine, Bethesda, Maryland, MS A17, p. 463. This manuscript is the third oldest Arabic medical manuscript known to be preserved today.* Adapted from: https://www.nlm.nih.gov/exhibition/islamic_medical/islamic_06.html.

*Figure 2.6 Latin translation of Al-Hawi.* Adapted from: http://www.muslimheritage.com/article/al-razi-small-pox-and-measles.

*Figure 2.7 European depiction of the Muhammad ibn Zakariya Al-Razi in one of Gerard of Cremona's Latin translations of his works of medicine (1250–60).* Adapted from: https://en.wikipedia.org/wiki/Muhammad_ibn_Zakariya_al-Razi.

in Bagdad, he ordered meat to be hung in various locations of this city and ultimately he selected the place where the effects of meat deterioration were least (Tadbakhsh, 2003; Ibn Abi Usaybi'ah, 2013). Ethics has always been an important concern for Islamic scholars, especially Razi. For example, Razi advised his students against speaking about deadly poisons in the presence of kings or merchants (Rāzī, 2008a,b). *Al-Mansūrī fī al-Ṭibb* is the most important work by Rāzī. In the eighth chapter of this book, small animal bites and various toxins have been described. One of the important things in this book is the different methods listed for repelling insects, snakes, animal bites, and beasts of prey. Furthermore, for every toxin, poisoning treatment has been provided (Rāzī, 2008a,b).

## 2.2.7 Avicenna (980–1037 AD)

Abū 'Alī al-Ḥusayn ibn Abd Allāh ibn Sīnā was a philosopher, physician, mathematician, and certainly the greatest scientist of the Islamic world and a shining light in history (Fig. 2.8). He is known as Avicenna in Western countries (Fig. 2.9) (Tadjbakhsh, 2003). Ibn Sina divided his *Canon of Medicine* (Figs. 2.10 and 2.11) into five books. The first book—the only one to have been translated into English— concerns basic medical and physiological principles as well as anatomy, regimen, and general therapeutic procedures. The second book is

*Figure 2.8 An imaginary drawing of Ibn Sina.* Adapted from: http://www.muslimheritage.com/article/al-razi-smallpox-and-measles.

*Figure 2.9 Commemorative medal issued by the UNESCO in 1980 to mark the 1000th birth anniversary of Ibn Sina.* Adapted from: http://www.1001inventions.com/node/1611.

*Figure 2.10 Images from Latin translation of Canon.* Adapted from: http://www.1001inventions.com/node/1611.

on medical substances, arranged alphabetically, following an essay on their general properties. The third book concerns the diagnosis and treatment of diseases specific to particular parts of the body, while the fourth covers more general conditions not specific to particular regions of the body, such as poisonous bites and obesity. The final, fifth book, is a formulary of compound remedies (Nasser et al., 2009). This fifth volume of the *Canon of Medicine* has a chapter with detailed descriptions of poisoning (Heydari et al., 2013). Avicenna discusses poisons

*Figure 2.11 An illustrated page of the Canon in a Hebrew translation.* Adapted from: http://www.1001inventions.com/node/1611.

and venoms. In addition to the treatment of poisoning, he suggests prevention. He even mentions drugs that can be effective in preventing poisoning and capable of neutralizing the effects of toxins, such as a mixture of fig, walnut, salt and rue (*Ruta graveolens*) leaves, or zedoary (*Curcuma zedoaria*) root. Diagnostic methods for poisoning via symptoms, body odor, and appearance of vomiting have also been highlighted in this book (Ibn Sina 2008, 2010a,b). Avicenna suggests both general treatment of poisoning and also a special antidote for each poison (Ibn Sina, 2007).

The toxic effects of opioids are also discussed in detail by Avicenna. He discusses the mechanisms of opium poisoning, clinical manifestations, and treatment against herbal poisons. Most of these toxic effects have been confirmed by current research. Avicenna mentions that these symptoms can have 2-day delay after opium ingestion. He also proposed 7 g as the lethal dose for opium (Heydari et al., 2013).

### 2.2.8 Jurjānī, Ismāʿīl ibn Ḥasan (1042−1137 AD)

Ismāʿīl ibn Ḥasan Al-Husaini, Zain Al Din Al Jurjani (Jorjani), one of the most famous Persian physicians and prolific writers on medicine of his time, wrote an important book "Zakhire Khūrazmshāhi." This book is an extensive encyclopedia of medical topics in general and in detail; diseases, symptoms, etiology, diagnosis, and treatment (Tadjbakhsh, 2003). Volume IX of this book is devoted to toxicology, and the author proposes subjects similar to those of Avicenna. He mentions that if a person wants to go to a place where it is possible to be poisoned, he can eat something to prevent poisoning, and for this purpose he suggests various foods and natural drugs. Jurjani discusses venomous snakes and their characteristics and appearance of slough, and also treatment of snake bites. Information on bites of insects, scorpions, spiders, and tarantulas, and their treatment are also available in this volume (Jorjani, 2001).

### 2.2.9 Ibn al-Tilmīdh, Hibat Allāh ibn Ṣāʿid (1073−1164 AD)

Ibn al-Tilmīdh was a Christian physician, pharmacist, and poet, who lived in Bagdad (Ibn Abi Usaybi'ah, 2013). He wrote a treatise on "empirical drugs" (muġarrabāt) around the middle of the 12th century. The two antidotes discussed, fārūq theriac and the viper pastilles, originated in Greek prototypes. Different Arabic versions belonged to the stock of the medieval Islamic pharmacy (Kahl, 2010).

### 2.2.10 Ibn Zuhr, al-Malik ibn Abībal-ʿAlāʾ (Abu Marwan) (1072−1162 AD)

Ibn Zuhr, known as Avenzoar in the West, was born into a famous Andalusian family of physicians at Seville. He learned medicine from his father. During the Almohad caliphate of Abd al-Mumin, Ibn Zuhr compiled his major book "Kitab al-Taysir fi al-Mudawat." This book includes a vast materia medica in 30 chapters. It covers both Ibn Zuhr's personal views and recollections of his exile in Morocco and case histories in the manner of those written by Razi. This book was translated into Hebrew and Latin and was an essential textbook in European universities during the 18th century. Avenzoar is thought to have been the first person to describe bezoar stones as medicinal items.

These stones are formed in the digestive organs of animals and were believed to have the power of a universal antidote against any poison. He wrote "*al-taryaq Al-sabien*" and dedicated it to the caliph, Abd al-Mumin (Byrne, 2012; Golzari et al., 2013; Ibn Abi Usaybi'ah, 2013).

### 2.2.11 Jamshīd Ghyāth al-Dīn bin Mas'ud bin Mohammed al-Kāshī (1380–429 AD)

Ghiyāth al-Dīn Jamshīd Mas'ūd al-Kāshī (or al-Kāshānī) was a masterful Persian astronomer who was active in the field of medicine. A book in the field of toxicology titled "Tadārok al-Somum" was attributed to him (Fig. 2.12) (Kāshani, 2008; Schmidl, 2014).

## 2.3 TOXICOLOGISTS AFTER 1500 AD

### 2.3.1 'Imād al-Dīn Shīrāzī, Mahmūd ibn Mas'ūd (1515–92 AD)

He has authored several books on toxicology, including "Resale-al-Sammie" (Fig. 2.13), "Marefat-al-somum" (Fig. 2.14), and "Resale dar Somum." The book "Resale dar Somum" contains definitions of food, drug, and poison and includes two chapters on the signs, symptoms, and treatment of poisoning. At the end of the book, the author implies that there are natural drugs that are harmful and that animals seek to escape from their smell. Imad al-Din had written this book during the Safavid era by the request of Mashhad ruler Morteza Quli Khan. He had written, as well, another popular manuscript, Resale Pādzahrie (Fig. 2.15). Physicians during the Safavid era discovered a powerful antidote that is naturally occurring; its Persian name is "pād-zahr" (Nami and Tavana, 2009; Elgood, 1996). Pād-zahr, known as bezoar in

*Figure 2.12 Tadarok al-Somum: written by Ghyās al-dyn Kāshani in the 14–15th century.* Adapted from: Library, Museum and Document Center of Iran Parliament http://ical.ir/.

*Figure 2.13 Resale-al-Sammie written by Imād al-Din Shirāzi in 16th century.* Adapted from: Library, Museum and Document Center of Iran Parliament: http://ical.ir/.

Latin, was the stone produced in the stomach of a goat with the scientific name of *Capra aegagrus* known as a Bezoar goat (Fig. 2.16). It was originally Pāt-zahr in ancient Persia. In the Sassanid Pahlavic language, the official language of Persia during the Sassanid dynasty (224–637 AD), pat means "anti" and zahr means "poison." Therefore, Pāt-zahr means anti poison or antidote (Zargaran et al., 2013; Duffin, 2013). The history of the use of bezoar goes back to many years before the

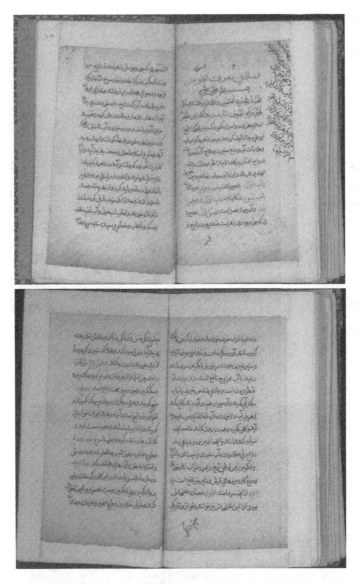

*Figure 2.14 Marefat-al-somum written by Imād al-Din Shirāzi in 16th century.* Adapted from: Library, Museum and Document Center of Iran Parliament: http://ical.ir/.

Safavid era, according to Haly Abbas (925–994 AD). Razi and Avicenna had also implied the existence of bezoar in their manuscripts. Soon, it became known in Europe. Nicolas-Hugues Ménard, a French Benedictine scholar, had written that bezoar had saved the life of the king of England. Queen Elizabeth-I always kept an amount of bezoar

*Figure 2.15 Resale Pādzahrie written by Imād al-Din Shirāzi in 16th century.* Adapted from: Library, Museum and Document Center of Iran Parliament: http://ical.ir/.

*Figure 2.16* Capra aegagrus *or Bezoar goat which a stone known as pād-zahr produced in its stomach.* Adapted from: https://commons.wikimedia.org/wiki/File:Carmel_Hai-Bar_-_Capra_aegagrus_creticus_%283%29.JPG.

under the signet of her ring. The pharmaceutical textbooks in England between 1618 and 1740 AD, had mentioned bezoar as a conventional drug. In 1808 AD, Fath-Ali Shāh, the second Qajar Shah of Persia, sent some bezoar as a valuable gift to Napoleon. Imad al-Din in his book, *Resale Pād-zahr*, introduced pād-zahr and implied that its counterfeits can be found in Syria. Furthermore, he described methods for distinguishing counterfeits from the original. Another important achievement of Imad al-Din was the innovation of new methods for detoxification of some poisons, especially mercury (Nami and Tavana, 2009; Elgood, 1996).

### 2.3.2 Mohammad Momen Tonkāboni (17th century, 1677)

Hakim Momen was the son of Mirza Mohammad Zamān Tonkāboni and the head of Persian physicians during the reign of Shah Soleimān in the Safavid era. He had written *Tohfe al-Momenin* based on his own experience and that of his father. A chapter of this book was dedicated to toxins and methods of detoxification. Some recommendations for prevention from poisoning were also recorded in that chapter. Recommendations included the prohibition of eating uncovered food that was exposed to insects, eating sour things that remained for some time in copper containers, and eating fruits, especially grapes, without washing. Also, some herbs, including *Pyrethrum roseum* seed and *Achillea millefolium* root were reported to have a preventive role in poisoning. Moreover, methods for the prevention and management of insect bites were mentioned (Tonkaboni, 2011).

### 2.3.3 Aghili (18th century)

Mohammad Hossein Aghili Alavi Khorasani Shirazi was one of the most popular Persian physicians and pharmacists during the Afsharid dynasty. In his book, *Makhzan al-Advia*, he introduced more than 1700 single drugs of natural origin (plant, mineral, and animal) along with their medicinal properties. Moreover, for toxic plants he described the symptoms of poisoning and methods of detoxification. For example, he cited hemlock (*Conium maculatum*) root and implied that about 6.4 g of that root can cause dementia, dizziness, blurred vision, hiccups, CNS depression, and coldness of the extremities and ultimately results in convulsions, asphyxia, and death. For reducing the toxicity of this herb, Aghili suggested that first it should be finely powdered, then macerated in milk for 3 days, dried and then mixed with almond oil for 1 week. After these procedures, it can be used more safely for management of pain and insomnia (Aghili, 2011).

### 2.3.4 Hakim Seyed bin Hasan bin Seyed Mahdi

Hakim Seyed bin Hasan bin Seyed Mahdi (19–20th century) wrote a book titled *"Taryāq Kabir."* Its subject is toxicology, management of poisonings, insect stings, rabid dog bites, and single and compound drugs used for treatment of these problems (Ibn Seyed Mahdi, 2004).

## 2.4 DISCUSSION

The science of toxicology achieved a unique position in Islamic countries. Many scientists from these countries contributed to the field and wrote prolifically on the subject, especially during 9th and 10th centuries. Different aspects of poisons, including their chemistry, pharmacology, and pharmacokinetics, have been considered in the literature. Medieval Islamic toxicologists have described identity, types, and constituents of poisons as well as their possible route of entry to the body, diagnostic methods for different types of poisons, and methods of detoxification. Not only natural and synthetic poisons, but also animal and insect bites and venoms were studied by medieval Islamic toxicologists. They focused on both treatment and prevention. A specific example is Avicenna's suggestion to combine the plants, *Ficus carica, Juglans regia*, and *R. graveolens* to ameliorate the effects of toxins. Moreover, he implied that *C. zedoaria* root is particularly efficacious in the detoxification of poison (Ibn Sina, 2010b). Rāzī in *Al-Mansūrī fi al- ibb* mentioned that one of the beneficial drugs for scorpion sting is about 5 g root of *Citrullus colocynthis* taken orally. Furthermore, he touched on the symptoms of oleander poisoning along with its treatment. Rāzī in that book discussed a four part antidote that is very effective for scorpion and reptile stings, and composed of *Gentiana lutea, Commiphora myrrha, Aristolochia*, and *Laurus nobilis* (Rāzī, 2008a,b).

Observing ethical issues in the field of toxicology and constructively critiquing the transmitted theories of earlier toxicologists are among other characteristics of medieval Islamic toxicologists. The study of the manuscripts of medieval Islamic toxicologists would be well rewarded by offering insights into their valuable clinical experiences and suggesting possible treatment protocols that should be explored.

## ACKNOWLEDGMENT

We would like to thank Dr. Mohammad Reza Shams-Ardekani for providing some books and literatures required for completing this chapter.

## REFERENCES

Aghili MH: In Shams-Ardekani MR, Rahimi R, Farjadmand F, editors: Makhzan al-*Advia*, Tehran, 2011, Tehran University of Medical Sciences, p. 56. 502.

Byrne JP: Encyclopedia of the Black Death, *ABC-CLIO.* V 1:33–34, 2012.

Davari H: Look at the etymology of some entries of Persian dictionary, *Spec Issue Acad* 7:236–240, 2013.

Duffin Cl: Porcupine stones, *Pharm Hist* 43:13–22, 2013.

Elgood S: In Javidan M, editor: *Medicine during Safavid era*, Tehran, 1996, University of Tehran, pp. 25–29.

Golzari SE, Khan ZH, Ghabili K, et al: Contributions of medieval Islamic physicians to the history of tracheostomy, *Anesth Analg* 116:1123–1132, 2013.

Heydari M, Hashempur MH, Zargaran A: Medicinal aspects opium as described in Avicenna's canon of medicine, *Acta Medico-Hist Adriat* 11:101–112, 2013.

Ibn Abi Usaybi'a: In Zaker M, editor: *Uyun al-anba' fi tabaqat al-atibba*, V2, Tehran, 2013, zaeem.702, 864–866, 1074–1165.

Ibn Seyed Mahdi SH: *Taryāq Kabir*, Tehran, 2004, Research Institute for Islamic and Complementary Medicine.

Ibn Sina H: In Shirazi F, editor: *Overview of canon*, Tehran, 2007, Research Institute for Islamic and Complementary Medicine, p. 116.

Ibn Sina H: In Ghazvini A, editor: *Health maintenance*, Tehran, 2008, Research Institute for Islamic and Complementary Medicine, p. 60.

Ibn Sina H: In Sharafkandi A, editor: *Canon of medicine*, V2, Tehran, 2010a, Soroush, p. 35, 50.

Ibn Sina H: In Sharafkandi A, editor: *Canon of medicine*, V7, Tehran, 2010b, Soroush, pp. 6–7.

Jorjani SE: In Tajbakhsh H, editor: *Zakhire khwarazmshahi*, Tehran, 2001, Tehran University of Medical Sciences, pp. 693–714.

Kahl O: Two antidotes from the 'empiricals' of Ibn at-Tilmīḏ: antidotes mentioned in an unpublished treatise on mujarrabāt empirical drugs composed in mid-12th century. Commentary, edition & translation, *J Semitic Stud* 55:479–496, 2010.

Kāshani Gh: *Tadāroke Somum*, Tehran, 2008, Research Institute for Islamic and Complementary Medicine.

Klein E: *A comprehensive etymological dictionary of the English language*, Amsterdam, 1971, Elsevier Scientific Publishing Co.

Kraus P: *"Jâbir ibn Hayyân: Contributions à l'*, Histoire des Ideés Scientifiques dans l'Islam, Le Caire, 1943, Impr. de l'Institut français d'archéologie orientale.

Levey M: Medieval arabic toxicology: the book on poisons of ibn Wahshīya and its relation to early Indian and Greek texts, *Trans Am Philos Soc* 56:1–130, 1966.

Mohagheghzadeh A, Morovvat MH, Ghasemi Y, et al: *An investigation in Iranian Ancient Medical Sciences, Article collections on the honor of Professor Dr. M.R. Mohareri*, Tehran, 2006, Iranian Academy of Medical Sciences Publications.

Mousavi Bojnourdi MK: *Great Islamic encyclopedia*, Tehran, 1988, Center of Great Islamic Encyclopedia.

Nafisi A: Surveys on symptoms and type of poisons used for poisoning of the Prophet and Shiite Imams, *J Fac Theology Islamic Educ* 11:1–24, 1974.

Nami V, Tavana F: *Collection of books specific to Research Institute for Islamic and Complementary Medicine*, Tehran, 2009, Research Institute for Islamic and Complementary Medicine.

Nasser M, Tibi A, Savage-Smith E: Ibn Sina's Canon of Medicine: 11th century rules for assessing the effects of drugs, *J Royal Soc Med* 102:78–80, 2009.

Rāzī AM: In Tafaghod R, editor: *Rhazes' treatise to one of his students in the field of medical ethics*, Tehrsn, 2008a, Tehran University of Medical Sciences, p. 40.

Rāzī AM: In Zaker M, editor: A*l-Mansūrī* fī al- ibb, Tehran, 2008b, Tehran University of Medical Sciences.

Schmidl PG: Kāshī: Ghiyāth (al-Milla wa-) al-Dīn Jamshīd ibn Mas ç ūd ibn Ma mūd al-Kāshī [al-Kāshānī], *In:The Biographical Encyclopedia of Astronomers.* Thomas Hockey,Virginia Trimble, Thomas R. Williams, Katherine Bracher, Richard A. Jarrell, Jordan D. Marché II, F. Jamil Ragep, JoAnn Palmeri, Marvin Bolt (eds.)., Springer Reference. New York: Springer, 2007, pp. 613–615 :1161–1164, 2014.

Sezgin F: *History of Arabic writings.* Translated by Publishing Institute of the Academy and Book House, Tehran, 2001, Ministry of Islamic Culture and Guidance, 322–328, 374, 384.

Shahîd I: *Byzantium and the Arabs in the Sixth Century*, Washington, 2010, Harvard University Press.

Tadjbakhsh H: *History of human and veterinary medicine in Iran*, Lyon, 2003, Fondation Mérieux, 301, 311, 317–332, 508.

Tonkaboni MM: *Tohfe al-Momenin*, Qom, 2011, Noure Vahy.

Tschanz DW: Medieval Islamic pharmacy development of a science and profession, *Aspetar Sport Med J* 2:616–621, 2013.

Weekley E: In Murray John, editor: *An etymological dictionary of modern English*, New York, United States, 1921, Dover Publications. reprint (1967).

Zargaran A, Yarmohammadi H, Mohagheghzadeh A: Origin of Bezoar or Pāt-zahr in Persian medicine, *Pharm Hist* 43:43, 2013.

# CHAPTER 3

# Maimonides' Book on Poisons and the Protection Against Lethal Drugs

**Gabriele Ferrario**
Cambridge University Library, Cambridge, United Kingdom

## 3.1 MAIMONIDES

Moses Maimonides (Arabic: Abū ʿImrān Mūsā ibn ʿUbayd Allāh ibn Maymūn; Hebrew: Moshe ben Maymun or simply "The RaMBaM") was born in Cordoba, Spain, in 1138[a]. His family left Andalusia in 1148, when the surge to power of the Berber dynasty of the Almohads led to worsening of the living conditions for religious minorities. They traveled through Southern Spain and Morocco and settled briefly in Palestine, before finding permanent residency in the city of Al-Fusṭāṭ (Old Cairo)—at the time a lively center of economic, commercial, and intellectual exchange. Maimonides authored crucial works on the religious basis of Judaism and its philosophical stance: His *Mishneh Torah* ("Repetition of the Law") is still regarded as one of the most authoritative systematizations of Jewish Law, while his *Moreh ha-Nevukhim* ("Guide for the perplexed") is studied among the fundamental texts of medieval philosophy and theology. Maimonides' religious works include a commentary on the Babylonian Talmud (the *Perush Mishnah*), one on the Palestinian Talmud (only extant in fragments), a systematization of the principles of the Jewish faith (*Sefer Ha-Miṣvot*), and a number of epistles and *responsa* to queries sent from Jewish communities near and far. Requests for legal opinion from communities as far away as Southern India, and the fact that Maimonides' Judaeo-Arabic works were almost immediately translated into Hebrew (thus allowing their access by non-Arabic speaking Jewish communities) reveals the spread of Maimonides' fame as a religious and

---

[a]The traditional date 1035 has been questioned on the basis of a note by Maimonides himself that stated that at the completion of his *Perush Mishnah*, finished in 1068, he was 30 (see Goitein, 1980, p.155; Leibowitz, 1976, pp. 75–76; Bos and McVaugh, 2009, pp. XVII–XVIII).

Toxicology in the Middle Ages and Renaissance. DOI: http://dx.doi.org/10.1016/B978-0-12-809554-6.00003-2

*Figure 3.1 Portrait of Moses Maimonides—U.S. National Library of Medicine, Bethesda, MD, USA.*

intellectual authority during his own lifetime. He partnered with his brother, David, in a successful trading business with Yemen and India and appears to have continued this business after the loss of his brother in a shipwreck in 1169 (Goitein, 1980). Maimonides' intellectual production was not limited to religious and philosophical themes, but embraced medicine, with contributions of lasting importance and influence. He was also a practicing physician to Muslim rulers and their entourages, and attended to the medical needs of his own community (Ben Sasson, 1991; Goitein, 1980; Kraemer, 2004).

## 3.2 MAIMONIDES MEDICAL WORKS AND MEDICAL PRACTICE

Eleven of the numerous medical works attributed to Maimonides are considered authentic. They deal with both general medical themes and specific medical problems, and make use of Greek and Arabic sources and considerations derived from his actual medical practice (Ackermann, 1986; Bos, 2008). Originally composed in Judaeo-Arabic—the variant of Arabic used by Jewish communities in Arabic-speaking countries—his major medical writings are

*Figure 3.2 Statue of Moses Maimonides in Cordoba, Jewish Quarter, Creative Commons CC0 1.0 Universal Public Domain Dedication.*

four[b]: (1) the *Commentary on Hippocrates' Aphorisms* (*Sharḥ Fuṣūl Abuqrāṭ*); (2) the *Book of the Medical Aphorisms* (*Kitāb al-Fuṣūl fī al-ṭibb*); (3) the *Compendia on the Books of Galen* (*Mukhtaṣarāt li-kutub Jālīnūs*); and (4) the *Book on Poisons and the Protection against Lethal Drugs* (*Kitāb al-Sumūm wa al-taḥarruz min al-adwiya al-qattāla*). Other medical works recognized as authentic are: (1) the *Book on Coitus* (*Kitāb fī al-jimāʿ*); (2) the *Book/Treatise on Regimen Sanitatis* (*Kitāb/Maqāla fī Tadbīr al-ṣiḥḥa*); (3) the *Treatise on the Explanation of the Causes of Symptoms* (*Maqāla fī bayān al-aʿraḍ wa-al-jawāb ʿanha*); (4) the *Commentary on the Names of Drugs*

[b]For a detailed bibliography on Maimonides' medical works, see Bos and McVaugh, 2009, p. 293, n. 7. and pp. 317−18.

(*Sharḥ al-asmā al-ʿuqqar*); (5) the *Epistle on Haemorrhoids* (*Risāla fī al-bawāsīr*); (6) the *Treatise on Asthma* (*Maqāla fī al-rabw*); and (7) the *Book on the Rules of the Practical Part of the Medical Art* (*Kitāb qawānīn al-juz' al-ʿamalī min ṣināʿa al-ṭibb*)[c].

Maimonides' medical works not only reveal his widespread knowledge of the Graeco-Roman medical tradition and of Arabo-Islamic medicine, but also the direct observation of medical cases that he was confronted with as court physician for Al-Qāḍi al-Fāḍil and after 1199, for the son of Saladin, Al-Malik al-Afḍal.

## 3.3 THE TREATISE ON POISONS AND THE PROTECTION AGAINST LETHAL DRUGS[d]

The idea of writing a toxicological treatise was suggested to Maimonides by the counselor of the Ayyubid caliph Salaḥ al-Dīn (the famous opponent of the Crusaders, Saladin), ʿAbd al-Raḥīm b. ʿAlī al-Baysānī, more commonly known as al-Qāḍi al-Fāḍil, toward the end of the 12th century. According to the introduction to the work, Al-Qāḍi al-Fāḍil had already commissioned toxicological works of Egyptian physicians, but in the work commissioned of Maimonides, he requested the focus be directed on poisons and antidotes readily available in Egypt. Moreover, the treatments presented should be easy to self-administer, without the intervention of a physician[e].

The title by which this work is commonly known, *Book on poisons and the protection against lethal drugs* (*Kitāb al-Sumūm wa al-taḥarruz min al-adwiya al-qattāla*), is not preserved in any of the extant manuscripts of the work. It became the commonly accepted title since it was used by the biobibliographer Ibn Abī Uṣaybiʿa (13th century) and in the works that used his *Generations of Physicians* as a source (Ibn Abī Uṣaybiʿa n.d., p. 583)[f].

[c]The *Kitāb qawānīn*, originally believed to be a copy of *On Asthma* (Steinschneider, 1893, p. 767), is preserved in only one manuscript (Ms Madrid 6019, ff. 109–123a) and was translated and edited by Bos and Langerman (2014).
[d]The most extensive and detailed study of *On Poisons*, together with the critical edition and English translation of the Arabic text, a critical edition of the Hebrew and Latin extant versions, is found in Bos and McVaugh (2009). I have extensively relied on this excellent work for composing this chapter.
[e]See Bos and McVaugh (2009, pp. 5–6).
[f]For the titles of *On Poisons* preserved in the extant mss, see Bos and McVaugh, 2009, p. XIX.

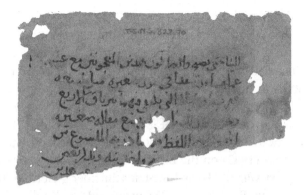

*Figure 3.3 Cambridge University Library, T-S NS 327.70 recto: A Genizah fragment from the introduction of Maimonides' On Poisons.* Reproduced by permission of the Syndics of Cambridge University Library.

It appears very likely that Maimonides originally drafted his medical works in Judaeo-Arabic (Arabic written in Hebrew letters), as his autographs found in the Cairo Genizah show, and they were only later transcribed in Arabic script for a wider audience (Blau, 1965, p. 41, n. 6; Hopkins, 2005, p. 90). The Arabic original of *On Poisons* survives in at least ten manuscripts, six in Arabic characters, and four in the Hebrew alphabet[g]: this fact, together with the presence of at least 16 manuscripts of the two Hebrew translations of the work produced by the end of the 13th century, testifies to the popularity of *On Poisons*, in both Jewish and non-Jewish circles.

A passage from the fifth chapter of the first part of *On Poisons* regarding the spiders known as *rutaylā'* (commonly considered to be spiders of the Tarantula family) is quoted in the zoological lexicon *Ḥayāt al-Ḥayawān*, a huge collection of prosaic and poetic quotations covering more than 900 animals mentioned in the Qur'ān, the Ḥadīth and popular proverbs, by the scholar of Islamic traditions Al-Damīrī (d. 1405)[h].

The fame of *On Poisons* continued in Europe, where three different Latin translations were produced starting from the early 13th century.

---

[g]For a thorough description of the Arabic and Judaeo-Arabic extant mss of *On Poisons*, see the detailed survey in Bos and McVaugh (2009, pp. XXI–XXIII).
[h]For a translation of Al-Damīrī's zoological work, see Jayakar (1906–08); the passage on *rutaylā'* is found at I:857. For Al-Damīrī's life see Kopf (1960–2007).

### 3.3.1 History of Studies

*On Poisons* has been the subject of numerous studies aimed at producing an authoritative edition of the Arabic or Hebrew texts, as well as translations into European languages. A first attempt at an edition of the Arabic text was undertaken by Hermann Kroner, whose efforts were cut short by his premature death, and are now preserved in manuscript form at the Countway Library of Medicine in Boston[i]. The Hebrew translation of the work conducted by Moses ibn Tibbon was edited by Süssman Muntner (1942) on the basis of a Parisian manuscript[j], and the same Hebrew text was the basis for Steinschneider's German translation (1873), from which Bragman (1926) produced an English translation. Further, English translations based on Muntner's edition were published by Muntner (1966) and by Rosner (1988). An early French translation based upon the Hebrew version of Ibn Tibbon was published by Rabbinowicz (1865). The definitive version of the *Book on Poisons* was completed by Bos and McVaugh (2009): it includes a thorough introduction to the text and the history of its translations and transmission, first critical editions of the Arabic text with an English translation, the critical edition of the two Hebrew translations and the three Latin translations of the work, a glossary of technical terms and *materia medica*, and an exhaustive bibliography.

### 3.3.2 Context and Sources

Interest in poisons and their remedies was widespread in antiquity and in the Hellenistic world. Monographs on this theme, such as the *Book on Poisons* by Rufus of Ephesus (2nd century), Galen's *De antidotis*, *De theriaca ad Pisonem liber* and *De theriaca ad Pamphilianum*, and a pseudo-Galenic *Book on Poisons*, expanded on the treatment of this topic, which was presented also in major Byzantine *compendia*, such as in the fifth book of Paul of Aegina's *Medical Compendium* (7th century). In the Arabo-Islamic world, poisons and antidotes entered medical encyclopedias by Muḥammad ibn Zakariyā Al-Rāzī, Ibn Al-Jazzār, Al-Mājūsī, and Ibn Sīnā (Avicenna). Maimonides' *Book on Poisons* found its place among other monographs on the same topic penned by medieval Islamic physicians and alchemists, like the *Kitāb al-Sumūmāt wa-dafʿ maḍārrihā* by Yaḥyā ibn Al-Biṭrīq (9th century),

---

[i]Boston, Countway Library of Medicine heb. 21, ff. 1–29; pp. 54–59.
[j]Paris, Bibliothèque Nationale héb. 1173.

the *Kitāb al-Sumūm* by Jābir ibn Ḥayyān (9th century), and the *Kitāb al-samāʿim* by Ibn al-Jazzār (10th century)[k.]

Maimonides explicitly references Arabic translations of Galen's *De antidotis* and *De simplicium medicamentorum temperamentis ac facultatibus* among the classics, and quotes remedies from Al-Rāzī's *Kitāb Manṣūrī fī al-ṭibb*, Pseudo-Rāzī's *Kitāb al-Fākhir* (Richter-Bernburg, 1983), the *Book on Theriac* by Ḥunayn ibn Isḥāq (9th century), Ibn Sīnā's *Canon of Medicine,* and Abū Marwān ibn Zuhr (12th century), whom Maimonides praises profusely as the most knowledgeable author when it comes to simple and compound drugs for curing poisoning. In a few passages of the treatise, Maimonides refers to "the elderly" or "the senior physicians" with whom he exchanged medical opinions and anecdotes and, in the discussion of the quality of antidotes for poisons, he explicitly relies on his own direct medical observations.

## 3.4 STRUCTURE AND CONTENTS OF *ON POISONS*[l]

*On Poisons* is divided in two main parts: The first is devoted to the treatment of bites of vermin and venomous animals, the second is concerned with the treatment of a person who has otherwise been exposed to poison.

The introduction of the work, after an extensive encomium for Al-Qāḍi al-Fāḍil, includes programmatic statements helpful for a general evaluation of the content of the work. The *Book on Poisons,* it states, is going to be brief and concise and will deal with the treatment that the poisoned person should receive promptly and the medications and foods he should take. However, it will not address the expensive and rare theriac and Mithridates' electuary—a medicinal preparation mixed with honey or a sweet substance[m]. Maimonides will choose from his sources only the simplest and most available remedies, easy to memorize and prepare, without the intervention of a physician, with instruction and doses.

---

[k]For a survey on toxicological works by Greek, Byzantine and Arab authors, see Ullmann, 1970, pp. 321–42.
[l]For this summary of the contents of the treatise, I rely on the Arabic critical edition and English translation in Bos and McVaugh (2009, pp. 1–62).
[m]For the composition of these two remedies, see Galen, *De Antidotis*, I:6–7 and II:1.

The section on the treatment of poisonous bites opens with a detailed explanation of the regimen to be followed: the bitten area should be tied and incised, and the poison should be sucked out by a healthy mouth (coated with olive oil or wine) or by way of cupping glasses. Vomit should be induced and the available theriac, electuary, compound, or simple remedy should be administered. The wound should be kept open to allow the poison to exit the body, and the bitten person should be kept awake to avoid the spreading of the poison enhanced by innate heat during sleep. Maimonides is clear in defining the limits of his treatise: if the patient's condition does not improve after eight hours, he should be referred to a physician, since it would thus be an extreme case, which falls outside the scope of his work.

In the following sections, Maimonides lists simple and compound topical remedies (pigeon, duck and goat dung, asafoetida[n], bdellium[o], etc.) to be applied to the bite; he then deals with remedies to be ingested with milk, vinegar, or water, if the poisoned person feels hot, or with wine or anise decoction, should they feel cold. Maimonides' attention to dosage is evident in his notes on the appropriate dose of medicament for younger patients and the influence of the seasons on the amount of remedy that patients can tolerate.

A chapter of the first part is devoted to remedies for the bite of specific animals that were commonly lethal in medieval Egypt: Scorpions, al-jarrārāt (a small, yellow, and extremely poisonous scorpion), rutaylā', bees and wasps, snakes, rabid dogs, and the possibly dangerous bite of nonvenomous animals that in certain conditions of bad temperament can lead to putrefaction and death. Maimonides is particularly cautious in warning against dog bites: one should be able to recognize mad dogs, since they walk alone, stumbling and keeping close to the walls, but in case of doubt, one should always take the caution of treating the bite of a dog as if it were rabid. The first part of the treatise concludes with a list of foods that are beneficial to bitten patients: tharīda (bread and soup, or omelette, and bread and meat), olive oil, clarified butter, fresh milk, figs, walnuts, hazelnuts, garlic, onion, river crabs (particularly for contrasting the bite of a rabid dog), and as much wine as the patient can tolerate.

[n]The dried latex obtained from the rhizome or root of various species of Ferula.
[o]The semitransparent resin extracted from the Indian bdellium tree (commiphora wightii) and the African myrrh (commiphora africana).

The second part of the treatise is devoted to poisons that are ingested unknowingly. It opens with a theoretical discussion of color, taste, and smell aimed at a very practical consideration: food that tastes and smells foul should be inspected carefully, since the presence of poisons in food is often revealed by a change in taste, smell and color, and assassins tend to choose strong tasting foods or drinks (like wine) or naturally bad smelling foods for masking their poisons.

Maimonides presents a detailed explanation of the regimen to be followed: the poisoned person should be forced to vomit and purge as many times as possible, ingesting oils, milk and fats to neutralize the effects of the poison. Simple and compound remedies useful in cases of poisoning are similar to the ones used for bites: emerald, bezoar, lemon seed, serpent root, animal rennet, the great theriac, Mithridates' electuary and the theriac of four ingredients[P].

The last chapter of *On Poisons* is devoted to various common poisonous substances and how to detect their presence in food or drink: mineral poisons (like litharge, verdigris, and arsenic) change the color of food; opium changes its smell; and milk of the latex plant and honey of the marking-nut change its taste. One should be very careful with hemlock, henbane, shells and seeds of mandragore, Spanish flies, mushrooms, truffles and black nightshade: these are sometimes very deadly and less detectable. Maimonides reports that he heard from senior physicians that adulterous women would mix menstrual blood taken at the beginning of their menses with their husbands' food, resulting in death or, as peculiar as it may sound, causing limbs to fall off due to suppurating elephantiasis.

## 3.5 THE HEBREW TRANSLATIONS AND THEIR CIRCULATION

As is common with other works by Maimonides, *On Poisons* received an early Hebrew translation by a member of the Tibbonides family, Moses ben Samuel ibn Tibbon (13th century), who provided translations of Arabic names of plants and remedies also into Latin and a Romance language, usually Occitan (Bos and McVaugh, 2009, p. XXVII). Ibn Tibbon's became by far the most famous of the Hebrew translations of

---

[P]The ingredients are listed by Maimonides himself in the opening of the fourth chapter of the first part of this work: myrrh, peeled laurel seed, Roman gentian and birthworth.

the treatise and the 12 known extant manuscript copies testify to its widespread fame[q]. Ibn Tibbon's translation was also used by later Jewish physicians: Shem Tov ibn Isaac, a 13th century Provençal physician, quotes *On Poisons* in a passage of his *Sefer ha-Shimmush* (Bos et al., 2011), a Hebrew translation of Al-Zahrāwī's *Kitāb al-taṣrīf*. The Hebrew version of *On Poisons* is moreover quoted by Levi Ben Abraham ben Ḥayyim (d. 1315) in his encyclopedic work *Livyat Ḥen*, by Jacob ben Ḥayyim Farissol, Solomon Vivas and Nethanel Caspi (three 15th century Provençal students of Solomon b. Menahem Frat Maimon), and by Don Vidal Joseph ibn Labi (15th century), author of a Hebrew translation of Joshua Lorki's (d. c. 1419) book on plants and herbs (Steinschneider, 1873, p. 40; Bos and McVaugh, 2009, pp. XIX–XX).

## 3.6 THE LATIN TRANSLATIONS AND THEIR CIRCULATION[r]

Maimonides' contribution to toxicology entered the Latin world through three different translations. The first one was produced at the beginning of the 14th century by Armengaud Blasie of Montpellier, nephew of the famous Arnold of Villanova, with the title *De Venenis* and a dedication to Pope Clement V. It is now preserved in four complete manuscripts datable to the 14th and 15th centuries, in which the organization of the chapters does not follow that of the original maimonidean text, possibly due to a rearranged *Vorlage*. The second Latin translation is preserved in six anonymous manuscripts. McVaugh (Bos and McVaugh, 2009, p. XL) has argued that this translation is likely to be the work of Giovanni da Capua, a converted Jew who worked as physician at the Papal court around 1300, and who is possibly also the author of the anonymous Latin translations of Maimonides' treatises *On Asthma* and *On Coitus* (Bos and McVaugh, 2002, 2:XXIII-XXXVI). This Latin translation was used as a source by the two French physicians Henry de Mondeville (d. 1313) and Guy de Chauliac (d. 1368), both authors of pioneering treatises on surgery. The third Latin translation is only preserved in a single Vatican manuscript (Ms Vatican Pal. Lat. 1146)[s]: the

---

[q]For a detailed descriptive list of the extant Hebrew mss, see Bos and McVaugh, 2009, pp. XXIV–XXIX.
[r]A detailed analysis of the extant Latin manuscripts of Maimonides' *On Poisons* and their relationships is found in Bos and McVaugh (2009, pp. XXXIII–LIII).
[s]Schuba (1981, pp. 101–2) believes this to be the first Latin translation by Armengaud.

manuscript preserves a completely different Latin translation from the previous two, characterized by a strict word-for-word correspondence with the Arabic text, though with a tendency to delete or summarize certain sections.

## REFERENCES

Ackermann H: Moses Maimonides (1135–1204): Ärtzliche Tätigkeit und medizinische Schriften, *Sudhoffs Archiv* 70(1):44–63, 1986.

Ben Sasson M: Maimonides in Egypt: the first stage, *Maimonidean Studies* 2:3–30, 1991.

Blau J: *The Emergence and Linguistic Background of Judaeo-Arabic: A Study of the Origins of Middle Arabic*, London, 1965, Oxford University Press.

Bos G: Maimonides' medical works and their contribution to his medical biography, *Maimonidean Studies* 5:244–248, 2008.

Bos G, Hussein M, Mensching G, et al: *Medical Synonym Lists From Medieval Provence: Shem Tov ben Isaac of Tortosa: Sefer ha-Shimmush. Book 29*, vol. 1, Leiden, 2011, Brill.

Bos G, Langerman YT: *On the Rules Regarding the Practical Part of the Medical Art*, Provo, Utah, 2014, Brigham Young University Press.

Bos G, McVaugh MR: *Maimonides, On Asthma*, Provo, Utah, 2002, Brigham Young University Press.

Boss G, McVaugh MR: *Maimonides, On Poisons and the Protection Against Lethal Drugs*, Provo, Utah, 2009, Brigham Young University Press.

Bragman LJ: Maimonides' treatise on poisons, *Medical Journal and Record* 124:103–107, 169–171, 1926.

Goitein SD: Maimonides, man of action—a revision of the master's biography in light of the Genizah documents. In Nahon C, Touati T, editors: *Hommage à Georges Vajda*, Louvain, 1980, Peeters, pp. 155–167.

Hopkins S: The languages of Maimonides. In Tamer G, editor: *The Trias of Maimonides: Jewish, Arabic and Ancient Culture of Knowledge*, New York, 2005, de Gruyter, pp. 85–106.

Ibn Abī Uṣaybiʿa: *ʿUyūn al-anbāʾ fī ṭabaqāt al-aṭibbā*, Beirut, n.d., Dār Maktabat al-Ḥayat.

Jayakar ASG: *Ad-Damīrī's Ḥayat al-ḥayawān (A Zoological Lexicon)*, London and Bombay, 1906–1908, Luzac & Co. and Taraporevala, Translated from the Arabic, 2 vv.

Kopf L: Al-Damīrī. In Bearman P, Bianquis T, Bosworth CE, Van Donzel E, Heinrichs WP, editors: *Encyclopaedia of Islam*, 2nd ed., Leiden, 1960–2007, Brill. [Consulted online on November 19, 2016 < http://dx.doi.org/10.1163/1573-3912_islam_SIM_1685 >; First published online: 2012].

Kraemer JL: Maimonides' intellectual milieu in Cairo. In Levy T, Rashed R, editors: *Maïmonide: Philosophe et savant (1138–1204)*, Leuven, 2004, Peeters, pp. 1–37.

Leibowitz JO: Maimonides: Der Mann und sein Werk: Formen der Weisheit, *Ariel* 40:73–89, 1976.

Muntner S: *Samme ha-mavet veha-refuʾot ke-negdam*, Jerusalem, 1942, Rubin Mass, Translated by Mosheh ibn Tibbon.

Muntner S: *Treatise on Poisons and Their Antidotes*, Philadelphia, PA, 1966, Lippincott.

Rabbinowicz IM: *Traité des poisons avec une table alphabétique de noms pharmaceutiques arabes et hébreux d'apres le Traité des synonymies de m. Clément-Mullet*, Paris, 1865, Librarie Lipschutz, (Reprint, 1935).

Richter-Bernburg L: Pseudo-Ṭābit, Pseudo-Rāzī, Yūḥannā ibn Sarābiyūn, *Der Islam* 60:48–77, 1983.

Rosner F: *Treatises on poisons, hemorrhoids, cohabitation*, Haifa, 1988, Maimonides Research Institute.

Schuba L: *Die medizinischen Handschriften der Codices Palatini Latini in der Vatikanischen Bibliothek*, Wiesbaden, 1981, Reichert.

Steinschneider M: Gifte und ihre Heilung: Eine Abhandlung des Moses Maimonides, auf Befehl des aegyptischen Wezirs (1198), *Archiv für Pathologische Anatomie und Physiologie und für Linische Medizin* 57:62–109, 1873. Reprinted in Sezgin F., editor: *Mūsā ibn Maymūn (Maimonides): Texts and Studies*, Frankfurt, 1996, Institute for the History of Arabic-Islamic Science.

Steinschneider M: *Die hebräischen Übersetzungen des Mittelalters und die Juden als Dolmetscher*, Graz, 1893, Akademische Druck- und Verlagsanstalt, (Reprint, 1956).

Ullmann M: *Die Medizin im Islam*, Leiden, 1970, Brill.

# CHAPTER 4

# Pietro d'Abano, *De venenis*: Reintroducing Greek Toxicology into Late Medieval Medicine

## Alain Touwaide

Institute for the Preservation of Medical Traditions, Washington, DC, United States

Sometime in the beginning of the 14th century, the physician, natural scientist, and philosopher Pietro d'Abano (c. 1250–1316), from the town of Abano in the vicinity of Padua, compiled a small book entitled in Latin, *De venenis* (*On poisons* and *On venoms*, since the Latin term designates both venom and poison as Latin does not have a specific term to distinguish these two categories of toxic agents). The treatise is divided in two major parts: the first is a general analysis of the concept of venom and poison (*venenum* in Latin), and the second contains a clinical description of the symptoms following dermal or oral exposure to venoms and poisons, together with methods for eliminating them from the organism or counteracting their effects. The substances under consideration (76 in total) come from the three natural kingdoms: Mineral (13 substances), vegetable (38 plants or parts of plants), and animal, including humans (25 venoms, animal products, human physiological liquids, and bites by humans).

The work can be read in as many as 73 manuscripts, of which at least 5 date to the 14th century. It was printed as early as 1472, together with another work of Pietro d'Abano, the *Conciliator litium medicinae*, and as a work in its own right as early as 1473, in three different editions, in Mantua and Padua; after another edition in 1474 in Rome, it was printed three times again in 1475, with an edition in Rome and two in Milan; later, it was printed in no less than 26 different editions from 1476 to 1579, from Venice in 1476 to Frankfurt in 1579 (Fig. 4.1). This intense printing activity did not preclude handwritten copy, since three such manuscripts were produced during the 16th century and one in 1586. Also, *De venenis* was translated into

Toxicology in the Middle Ages and Renaissance. DOI: http://dx.doi.org/10.1016/B978-0-12-809554-6.00004-4

*Figure 4.1 Portrait of Pietro d'Abano by Joost van Wassenhove (aka Giusto di Gand) (c. 1430–80) and Pedro Berruguete (c. 1450–1504), portrait of Pietro d'Abano (c. 1472/75) made for the studio of Federico da Montefeltro at Urbino, now in Paris, Louvres.*

several vernacular languages (French, Italian including Venetian, and Castilian) in the 16th century, with a printed edition of a French translation at the close of the 16th century (1593).

The dissemination of this small, and even modest, treatise is surprising, particularly when the work is compared to much more substantial and voluminous compilations of Pietro d'Abano, the most important of which is his *Conciliator litium medicinae*, mentioned above (usually identified as the *Conciliator*), originally written in 1303 with a revised version in 1310 (Fig. 4.2). This monumental compilation (the title of which can be translated as *Conciliation of debated medical questions*) analyzes in great detail 210 medical questions on which contemporary physicians diverged. The topics of these questions were not details related to the practice of medicine, but fundamental questions such as the methodology of science, nature, and human physiology, to mention just a few topics, as well as the role of physicians (defined as "ministers of nature"). The *Conciliator* is not only a report on divergent and controversial scientific and philosophical questions of

*Figure 4.2 Portrait of Pietro d'Abano, probably extracted from the edition of his* Conciliator differentiarum phi-losophorum et medicorum, *publish in Venice in 1496 by Boneto Locatello, or work by Francesco Bertelli.*

that time, but also a sum of knowledge in medicine and science, and an attempt to establish a path beyond doctrinal differences, providing medicine and science with a solid epistemological ground work for new developments to come.

A proper understanding of the apparently strange, if not unexpected and paradoxical fate of Pietro d'Abano's treatise *De venenis*, needs to take into consideration the larger context of science in Pietro's time, together with his own activity and his interest in ancient Greek medical, scientific, and philosophical literature (Fig. 4.3). During Pietro d'Abano's lifetime, science and medicine were flourishing. Greek scientific and medical treatises, including but not limited to those of Hippocrates (460—between 375 and 351 BC), Aristotle (384—322 BC), Dioscorides (1st century AD), Galen (129—after 216 AD), and others circulated all around the Mediterranean, had been translated into Arabic as early as the 9th century AD. They were used by Arabic scientists and physicians [such as ibn Sina (980—1037 AD), better known in the West as Avicenna, ibn Rushd (1125—1198), identified in

*Figure 4.3 Pietro d'Abano's treatise "De venenis" in the edition of Rome, 1490.*

the Western medieval world as Averroes, or Moses Maimonides (1135–1204)], and subsequently generated new insights and knowledge. Arabic scientific texts in turn were translated into Latin in the West from the late 11th century on, and circulated widely among medieval scientists, contributing to the development of new theories and works. The multiplicity of interpretations of scientific questions, major and minor, and their comparison resulting from their accumulation in an age of intense circulation of information, in some cases, revealed contradictions that might have inhibited scientific activity. In the *Conciliator*, Pietro d'Abano sought to eliminate the obstacles to science that could result from differing and sometimes conflicting theories.

Indeed, Pietro d'Abano went further and returned to the roots of subsequent science in order to resolve the possible contradictions between scientists from Ancient Greece to his time. At some point, he traveled to Constantinople, the capital of the Byzantine Empire, in order to learn Greek. Whether through good fortune or good connections, he was able to gain access to the most important scientific center in Constantinople in the 13th and 14th centuries, the *Hospital of King Milutin* (*Xenodochion tou Krali*, in Greek). It was built by the king of Serbia Stefan Uros Milutin II (c. 1253–1321), who married Simonida (c. 1294–after 1336), the daughter of the Byzantine Emperor Andronikos II Palaiologos (1259–1332). The Hospital (the exact location of which has not yet been archaeologically ascertained) was likely a vast complex not limited to just a hospital, but also including a library (probably the richest one in Constantinople at that time) and a school (which was not limited to medicine, but covered all sectors of knowledge and was granted the status of university, the last one of the Byzantine Empire at that time). There, Pietro d'Abano had access to a multitude of Greek texts, precisely those which had been translated into Arabic in the 9th century AD and were the basis of the scholarly study and elaboration by Arabic scientists.

Among the many texts, he found in this library were the *Problemata* (*Problems*) of Aristotle's school, the Lycaeum, presented in a question and answer format, in which the scientists of the Lycaeum sometimes contradicted each other in free intellectual exchanges. Pietro d'Abano could also have read works by Galen, such as the small treatise *De sectis* (*On medical schools*), an introduction to medicine exposing the approach to epistemology proposed by the Hellenistic and Roman era schools of medicine, and the major *De methodo medendi* (*On the therapeutic method*), which is a vast theoretical treatise on how physicians should approach, understand, and treat diseases. In addition, he had access to the major pharmaceutical treatise of antiquity, *De materia medica* by Dioscorides, which he read in two different textual versions.

Pietro left Constantinople at an unspecified date, perhaps when he felt he knew enough Greek to study independently the works above and certainly some others, and when he also had copies of these and other texts that he could take home with him. He sailed back to Padua and either commenced or returned to teaching at the university. At home, or while already in Constantinople, he probably translated into Latin all the texts he had assembled there. He was not the first to do so, as other scholars had worked on this as early as the 11th century, starting with the bishop of Salerno Alfano (d. 1085), who rendered into Latin the work on physical anthropology *De natura hominis* (*On the nature of man*) by the Byzantine bishop of Emesa Nemesius (4th century AD). Later, a judge, Burgundio of Pisa (c. 1110–1193), embarked on a vast program of translation of Greek works of all kinds (from theology to medicine) that included several treatises by Galen. Closer to Pietro d'Abano's time, a certain Niccolo (c. 1280–1350) from Reggio in Calabria, translated into Latin multiple other Greek scientific works, from Hippocrates to Galen. Contrary to his predecessors and contemporaries, however, Pietro d'Abano did not limit his activity to simply translating into Latin the Greek works that he brought back from the East. Instead, he integrated them into his teaching activity at Padua university, with the possible exception of the Galenic treatises which he may have rendered into Latin for his personal use. One such course focused on the *Problemata* of the Aristotelian school, and another on Dioscorides' *De materia medica*. In the latter, for example, Pietro often referred to the Greek text, which he had translated, and commented on, according to the didactic method of his time, noting, for example, differences between the two versions of the text he procured, commenting on the meaning of technical terms, and identifying, for example, the plants mentioned in the treatise.

Any interpretation of Pietro d'Abano's *De venenis* needs to be situated in the context of the above. A close examination of the work reveals that it is in large part a translation and adaptation of two Greek treatises on the same topic attributed to, but certainly not by, Dioscorides, author of *De materia medica*, which Pietro d'Abano discovered in Constantinople, in the same or in different manuscripts as the two treatises also circulated independently. The two treatises are traditionally identified by the same titles as the poems by Nicander of Colophon (2nd [?] century BC), although they are not related to the poems: *Alexipharmaka* (*On poisons*) and *Theriaka* (*On venomous animals*). The comparison of the texts is significant.

In Pietro d'Abano's chapter on the toxic effects of the ingestion of coriander (*Coriandrum sativum* L.), the symptoms of the intoxication are the following in the two treatises (the original text in Latin of Pietro d'Abano and in Greek of the Pseudo-Dioscorides, respectively, is followed in each case by an English translation which is mine):

### Pietro d'Abano, De suco coriandri
*Ille cui sucus coriandri datus fuerit patietur quasi destructionem intellectus, ac sicut ebrius videatur et tandem moritur stupide.*
#### On coriander juice
He who has been given coriander juice, suffers from an almost destruction of his intellect, seems as though drunk, and finally dies struck senseless.
### Pseudo-Dioscorides, Alexipharmaka, 9
τὸ δὲ κόριον ... μανίαν ἐπιφέρει τοῖς διὰ μέθην ὁμοίαν ...
Coriander ... provokes a madness similar to that of drunkenness ...

Similarly, the chapter devoted to cursed crowfoot (*Ranunculus sceleratus* L.) describes the so-called sardonic laughter (which is in fact a labiofacial paralysis that might be compared to a smile) in terms very similar to those of the pseudo-dioscoridean treatise:

**Pietro d'Abano, *De apio risus***
*Ille cui datum fuerit apium risus in potu, facit hominem extra mentem et continue ridet, propter hoc vocatur apium risus.*
**On the sardonic laugh**
He, who will be given a draught of cursed crowfoot, goes out of his mind and laughs continuously, and this is why it is called the sardonic laughter.
**Pseudo-Dioscorides, *Alexipharmaka*, 14**
ἡ δὲ σαρδόνιον λεγομένη πόα ... παραφορὰν διανοίας ... ἐπιφέρει καὶ σπασμὸς μετὰ συνολκὸς χειλέων ὥστε γέλωτος φαντασίαν παρέχειν ἀφ' ἧς διαθέσεως καὶ ἡ σαρδώνιος γέλως οὐκ εὐφήμως καθωμίληται ἐν τῷ βίῳ...
The plant called Sardonion ... provokes a transport of the mind and spasms with a contraction of the lips such as to give the impression of laughing, a disposition which has rightly been called sardonic laughter in daily life ...

The same process of translation/adaptation of the Greek text can be perceived in the chapters of *De venenis* devoted to venoms, when compared to the text of *Theriaka* attributed to Dioscorides. A significant case is the method to be used when a patient has been bitten or stung by a venomous animal.

**Pietro d'Abano, De morsibus aut puncturis ab aliquibus animalibus venenosis**
*Si vero aliquis morsus fuerit aut punctus ab aliquo animali venenoso et ignoret cujus sit punctura, tunc stringatur locus puncturae et ponantur super ipsum ventosae cum scarificatione, et sugatur locus cum ore servorum, ...*
**On the bites or stings by some venomous animals**
If anyone is bitten or stung by a venomous animal and ignores which animal stung, then it is necessary to tighten the place of the sting, to put cups upon it together with cuts, and to have the part of the body sucked by a slave ...
**Pseudo-Dioscorides, *Thêriaka*, *praef.* 15–16**
... ἐπὶ δὲ τῶν ἰοβόλων διὰ κατασχασμοῦ καὶ σικυῶν προσβολῆς, ἐκμυζήσεως καὶ περισαρκιμοῦ, τότε δὲ καὶ δι' ἀκρωτηριασμοῦ ...

... on venomous animals, (it is necessary to) scarify and apply cups, sucking and cutting the flesh (around the point that has been bitten or stung) and, in some cases, even to amputate ...

Being the last example, the case of the sea slug (*Aplysia depilans* Gmelin 1791) is described in the same way in both works, with the same symptoms:

**Pietro d'Abano, *De lepore marino aut rana marina***
*Ille cui lepus marinus aut rana marina data fuerit in potu, habebit vomitum unctuosum et totum ejus corpus efficietur turgidum et inflatum sicut ipposarca et fetebit eius anhelitus ...*
**On the sea hare or sea frog**
He who will be given the sea hare or sea frog in a draught will have a greasy vomit and all his body will be congested and swollen as if in dropsy, and his breath will be fetid ...
**Pseudo-Dioscorides, *Alexipharmaka*, 28**
.. ἔμετος χολώδης ... οἴδημα σώματος ... ὀδωδέναι τὸ στόμα ...
... a bilious vomiting (will follow) ... a swelling of the body ... breath will smell ...

As can be seen from these few examples, Pietro d'Abano's Latin text adheres closely to the Greek, to such an extent that we can almost consider his text a translation of the Greek, although he did not render the Greek original in a word for word translation, but expanded on the original text in what could almost be considered a paraphrase.

Interestingly enough, the two short treatises *On venoms* and *On poisons*, attributed to Dioscorides that were Pietro d'Abano's source, appear in some of the manuscripts of *De materia medica* that were in the library of the Hospital of the King in Constantinople when Pietro d'Abano sojourned in the city. It is thus highly probable that Pietro d'Abano consulted these manuscripts and brought back a copy of the two small treatises on toxicology, exactly as he did for *De materia medica*. Contrary to what he did with *De materia medica*, Pietro d'Abano does not seem to have based any teaching at Padua on the two works on toxicology.

Thus, *De venenis*, rather than being a wholly original construct, is rather the first translation into Latin of the two treatises *On venom* and *On poisons* attributed to Dioscorides, which had been unknown until then in the West. Not that there were no treatises on toxicology in previous medieval medical literature, the fact is that such previous literature resulted from a long tradition of accumulated knowledge that was continuously expanded, and that sometimes amalgamated material of popular origin, fantastic in nature as, for example, the famous tale of the Poisonous Virgin. Nothing of that type appears in Pietro d'Abano's treatise, which clearly aims to bring toxicology back to a solid footing, basing it upon knowledge coming directly from antiquity and Greek medicine.

Returning to what seemed to be the unexpected if not unexplainable success of the modest *De venenis*, particularly when compared to the *Conciliator*, we are now in a better position to understand its extraordinary dissemination. *De venenis* is not different in intention from the *Conciliator*: both aimed to rebuild science on a new basis and to provide it with a solid theoretical foundation. Between the two treatises, there was a difference, however, and a fundamental one: whereas the *Conciliator* refounded science by collecting all theories, comparing and confronting them, trying to go beyond their differences in order to extract a common ground that could be accepted by all scientists, whatever their philosophical and methodological background, *De venenis* did not engage into a discussion of current theories, but simply and more radically omitted current knowledge to replace it with the results of centuries of experience attained in the Greek world as codified by the two small treatises attributed to Dioscorides. This attribution was in a certain sense a guarantee of the quality of the information contained in the treatises, as Dioscorides had been unanimously recognized by physicians and scientists from the 1st century AD to Pietro d'Abano's time as the incontestable, if not unique authority on materia medica.

Historiography has attributed the origins of the Renaissance (allegedly characterized by a return to classical antiquity) and the subsequent birth of modern science to the fall of Constantinople in 1453, and the emigration of Byzantine scholars and scientists to the West. Pietro d'Abano's *De venenis* in the 14th century can be considered a precursor as it was widely consulted throughout Europe well until the end of the 16th century and opened new avenues for toxicological study and research. If his *De venenis* subsequently disappeared from scientific publications, it is not because it was abandoned, but probably, and more simply, because, thanks to its solid foundation on ancient Greek science, it contributed to the development of a new evidence-based toxicology, thus consigning it to the not insignificant role of historical influence.

## FURTHER READINGS

The first part of the Latin text of *De venenis has been published by Alberico Benedicenti, Pietro d'Abano (1250–1316)*. *Il Trattato "De Venensis"* (Biblioteca della "Rivista delle scienze mediche e naturali" 2), Florence, 1949, Leo S. Olschki.

On the treatise more specifically, see Alain Touwaide: Pietro d'Abano sui veleni. Tradizione medievale e fonti greche, *Med. nei Secoli* 20:591–605, 2008, which demonstrates that it is a translation of Greek texts.

More recently, see the study by Fredrick Gibbs, *Poison, Medicine and Disease in Late Medieval and Early Modern Europe* (Medicine in the Medieval Mediterranean 9). London and New York, 2017 Routledge, Chapter 3, which studies the theoretical concept of *venenum* in Pietro d'Abano's treatise.

# CHAPTER 5

# The Case Against the Borgias: Motive, Opportunity, and Means

Cathy Cobb

University of South Carolina, Aiken, SC, United States

## 5.1 INTRODUCTION

The Borgias, the powerful European Renaissance theocratic dynasty headed by Pope Alexander VI, has been accused of being covert "notorious...poisoners" who "dispatched...their rivals with a secret poison (Lane, 2014)." However, recent scholarship has questioned this verdict and asserted instead the Borgias acted overtly and appropriately—given the times—and although they were no strangers to subterfuge or assassination, they were innocent of the charge of poisoning (Noel, 2016a; Meyer, 2013).

Let us revisit the scene.

Originating around the 13th century in the town of Borja, in the Spanish region of Aragon, the devout Borgia family ultimately produced two popes; numerous bishops and cardinals; and even a saint (Batllori, 2003b; De Dalmases, 2003). Nonetheless, the first Borgia Pope, Alfonso de Borja, Callixtus III (1378−1458), already displayed the ambiguous moral philosophy that would plague the Borgia clan. To his credit, he ordered the retrial and posthumous vindication of Joan of Arc (Lynch, 2002); yet he issued a bull which strengthened Portugal's ability to enslave non-Christians (Thomas, 1999) and he engaged in unabashed nepotism. Callixtus appointed his nephew, Rodrigo Borgia (1431−1503), protonotary apostolic, which led the young man on the path to becoming a cardinal, and, eventually, Pope Alexander VI (Batllori, 2003b).

By the time of Rodrigo Borgia's election as pope (1492), the Great Schism had been over for 70 years, the Reformation was 30 years in the future, and Rome had recovered somewhat from the cesspool it had sunk into during the 14th century. The bishops of Rome enjoyed

Toxicology in the Middle Ages and Renaissance. DOI: http://dx.doi.org/10.1016/B978-0-12-809554-6.00005-6

relative respect, security, and comfort from benefices, indulgences, and the solid tax base of Rome, a major tourist attraction (Hibbert, 1994).

Papal life was good—but not perfect. The role of pope had evolved. Popes, in addition to their spiritual responsibilities, had become the secular heads of a nebulous entity called the Papal States. Popes now signed treaties, raised armies, and negotiated with warlords, in an effort to defend against outside intervention (Meyer, 2014a).

Of these warlords, there was one persistent burr under the papal saddle: Cardinal Giuliano della Rovere. Alexander had bested della Rovere in the contest for pope and continued to better him in papal power machinations. Della Rovere carried an unrelenting grudge for these injuries (Burchard, 2015a) and, as we will see later in this story, ultimately extracted his revenge in an enduring way. Otherwise, as skillfully as Michelangelo daubed paint from his palette, Alexander used friends, family, and influence to keep the Papal States secure.

But did he use poison?

Let's look at some of the allegations and examine them in light of motive, opportunity, and means.

## 5.2 ACCUSATIONS

It has been alleged that Pope Alexander VI or his agents, including his son Cesare and his daughter Lucrezia, poisoned to death, on or about

February 26, 1495, Turkish hostage, Djem, also spelled Ziem, Cem, or Zizim (Burchard, 2015b; Mathew, 1912a);
January 8, 1497, prisoner, Cardinal Virginio Orsini (Burchard, 2015c; Mathew, 1912b);
July 5, 1500, prisoner, Giacomo Caetani (Scarsbrook, 2011b);
December 1498–January 1500, Cardinal Juan de Borja Lanzol de Romaní (Mathew, 1912c; Lucas-Dubreton, 1954a);
June–August 1501, Cardinal Battista Ferrari of Capua (Mathew, 1912d; Lucas-Dubreton, 1954b); and
April 10, 1503, Cardinal Giovanni Michieli (Cloulas, 1989a).

## 5.3 MOTIVE

As we said, raising armies was among the Pope's concerns, and soldiers are seldom so pious that they do not require pay, so one strong motive would have been money. Indeed, avarice was alleged in the deaths of the cardinals: their benefices and properties reverted to the Church when they died (Mallett, 1981a). Political expedience was seen as the motive for others: troublesome warlords were not necessarily silenced by imprisonment and there was always the risk of rescue or escape. Public executions were likely to enflame their partisans, but quiet poisoning might avoid a vendetta. Beyond the above motives, revenge was cited, or simple pique.

## 5.4 OPPORTUNITY

Opportunities were abundant. Given that one's livelihood depended on the whim of the Pope, it is doubtful anyone would turn down an invitation to dine at his table. Moreover, detentions, arrests, and imprisonment were at the Pope's discretion, and substance intake for prisoners is easily controlled.

## 5.5 MEANS

They had the materials for poisons including

- animal [blister beetle (Cobb et al., 2013b), snakes, toads, salamanders (Oehme et al., 1975)];
- vegetable [mandrake, monkshood (aconite), nightshades (including belladonna, henbane), strychnine, hemlock (Thompson, 1993; Mann, 1992)]; and
- mineral [arsenic (Cobb et al., 2013c), corrosive sublimate (Patai, 1994a)].

But were these poisons good murder weapons? Probably not. A viable murder weapon should be deceptive, effective, and reliable, and in the 16th century, these attributes were not always available in poisons.

*Deceptiveness.* In a murder, the victim must be taken by surprise and the perpetrator unapparent—which is where poisons would seem to have the advantage. However, herbal poisons often have strong characteristic odor and taste and can exhibit diagnostic symptoms [such as the facial "death mask" caused by strychnine (Stevens, 1990a)] that could confound subterfuge. The lethal form of arsenic, arsenic trioxide, has a garlicky odor (Stevens, 1990b). Although this odor could be disguised as seasoning, a sprinkling of white crystals in one's food might raise suspicions. Given the solubility of arsenic trioxide in water,

17 g/L, and its LD50, 14.6 mg/kg (oral rat), a 70 kg victim would have to drink around 50 mL of saturated solution or eat about a gram of solid to have a reasonable chance of dying (Norman, 1977). This much solution or solid might noticeably alter the taste or texture of a food or beverage meant as a conveyance and possibly defeat an intended subterfuge. Arsenic trioxide is more soluble in caustic solution (Cobb et al., 2013a), but wine tends to be acidic rather than basic, so addition of such a solution to water or wine would probably cause the arsenic to precipitate, which would reduce the intended dosage. Moreover, wine with white flecks in the dregs would no doubt inspire reticence in the imbiber.

*Effectiveness.* In effectiveness, the European Renaissance poisons rate an A+. Their herbal poisons were sometimes too effective. Monkshood could act almost instantaneously, with characteristic symptoms (Stevens and Bannon, 2007), which would lower its deceptiveness rating (see above).

Nonetheless, slow-acting arsenic had its problems, too. In the 16th century, they knew to induce vomiting in cases of suspected poisoning and also how to cause diarrhea (Patai, 1994b). So, an ideal poison would have to work slowly enough for the assailant to gracefully exit the room, yet be strong enough to thwart countermeasures. There are legends that the Borgias had a secret poison called Cantarella, which was said to be

> ... *a brilliant white, slow acting venom, that was pleasant to the taste [and that] did not overwhelm a victim's vital forces by a sudden, energetic action. Instead, it worked to insensibly penetrate the veins, with a slow but deadly effect. (Scarsbrook, 2011a)*

Unfortunately, we have no information on the composition of this poison; therefore, we cannot admit it into evidence. In addition, to engineer such a poison consistently would require fairly sophisticated techniques—which brings us to the downfall of the Renaissance poisoner: reliability.

*Reliability.* The strength of the active principle of an herb could depend on time of year or time of day harvested (Cobb and Fetterolf, 2010). In addition, the amount of active ingredient extracted could depend on the method of preparation. If a solution were steeped in an attempt to increase the potency, the delicate organic compounds of the active agents could be destroyed (Cloulas, 1989c), which again would inadvertently reduce the dosage. Treatments to disguise odor or taste might likewise result in a reaction that denatured the active component. At the very least, stored herbal concoctions could be rendered impotent by simple rot.

Mineral preparations might not fare better. Cesare supposedly processed his arsenic to make it more soluble by force feeding it to a boar and then collecting the excrements (Scarsbrook, 2011c). However, it has been shown that organic forms of arsenic, such as those that might result from digestion by a pig, could be less toxic than the inorganic arsenic it had been fed (Burford et al., 2011), which again would reduce the effective dosage.

Furthermore, we now know individuals have different reactions to poisons. Not everyone is allergic to poison ivy (https://www.uihealthcare.org/content.aspx?id=236459), and a peanut can be a treat to one and poison to another. The state of the health of the individual could change their response to poison, as well as the frequency and the timing of the dosage (Eaton and Gilbert, 2013). All this would serve to make poisons very unpredictable tools for Renaissance would-be poisoners.

But enlightenment would come, and when it did, it came from a troubled, wandering, visionary—who was possibly the most brilliant instinctual healer, observer of human nature, and student of the cause and cure of disease that the age produced—Paracelsus (1493–1541) (Cobb, Goldwhite). He said,

*All substances are poisons; there is none which is not a poison. The right dose differentiates a poison and a remedy (von Hohenheim, 1975).*

The notion was still far removed from modern dose/response curves, but if the Borgias had better understood the factors affecting dosage, and how to control them, they might have made a reliable, effective, and deceptive tool for their arsenal—but they did not, so they could not.

The above arguments against the charge of poisoning (let alone murder) add to the plausibility arguments offered by others:

• Dead hostages are useless. The poisoning of the hostage Djem would have been a financial loss for Alexander (Scarsbrook, 2011d; Meyer, 2014b).

- Alive, the prisoner Virginio Orsini could have been ransomed (Meyer, 2014c; Mathew, 1912e). Dead, he could not. The same goes for Giacomo Caetani. His mother protested his so-called poisoning, so his family still wanted him alive.
- Cesare is said to have poisoned Cardinal Juan Borgia out of anger because the legate brought him bad news (Mathew, 1912f). Yet, if Cesare had jumped up and stabbed him, that might have some small bit of believability, but waiting to deviously poison him does not sound like a crime of momentary passion.
- The motive for killing the other cardinals was said to be financial gain (Partner, 1980; Cloulas, 1989a). The Pope giveth, the Pope taketh away. The estate of Cardinal Michiel was said to net the Pope 150,000 ducats—but was insufficient to fund the current military campaign. So Alexander created nine new cardinals, fetching 120,000 to 130,000 ducats each. Therefore, it would seem one new cardinal approximately equaled one dead one, so it might seem safer to add one more cardinal than to murder another (Scarsbrook, 2011e; Cloulas, 1989d).

So now, we are facing with the big question: Given the implausibility of the tales and the difficulties of the means, why do the Borgias still carry the onus of evil and corrupt poisoners?

For evil, Alexander VI's immediate predecessor, Innocent VIII, might be considered worse. He sanctified the horrific witch hunts that extended into the 18th century (Cobb, 2002). Alexander's penultimate predecessor, Sixtus IV, issued the papal bull that established the Spanish Inquisition (Mathew, 1912g). By contrast, Jews thrived in the shade of Alexander's Vatican. When Jews were fleeing from persecution, he opened the gates of the Papal States as a refuge (Carroll, 2002).

When the body of his eldest son was fished out of the Tiber, Alexander experienced deep grief and responded with a repentance that involved the entire Catholic Church. He ordered an ambitious top-down reform that touched on the excesses and assumed privileges of the churchmen. Unfortunately, as Mallet points out, the success of his reforms relied on the actions of the people he was attempting to reform, but at least he made the effort (Mallett, 1981b).

Then, was the Borgias' evil reputation based on his flagrant display of sexual immorality? Probably not. Celibacy for priests was only introduced in the mid-12th century, as a measure to reduce nepotism (http://historynewsnetwork.org/article/696). It was not unusual for popes to have children, and the use of women and boys for sexual purposes was acknowledged by the efforts to curtail the practice (Cloulas, 1989b). So considering all the above, the Borgias' behavior may have been a little ho-hum.

So why the reputation? Well, remember Cardinal Giuliano della Rovere, the person Alexander bested for pope? When della Rovere finally became pope, Pope Julius II, he tortured Alexander's servants until they confessed to Alexander's crimes, including the practice of poisoning (Cloulas, 1989e; Lucas-Dubreton, 1954c; Noel, 2016b).

But this evidence is suspect. Under torture, people will say what their interrogator wants them to say, which is why torture doesn't work (O'Mara, 2015).

So what is our verdict for the Borgias? Acquittal—with an opinion.

While there has been a resurgence of Borgia apologists, we can't let the pendulum swing too far the other way (Batllori, 2003a). There are scholars who see reasonableness in the historical reports of use of poisons by the Borgias (Scott and Sullivan, 1994). In addition, if a reliable, effective, and deceptive poison had been available, they would have most likely used it. As it stands, they were guilty of hedonistic excess, executions without due process, duplicitous politics, and greed. . .

But not poisoning.

## REFERENCES

Batllori M: *Alexander VI, Pope*, 2nd ed., *New Catholic encyclopedia*, 1, Detroit, 2003a, Gale, *Gale Virtual Reference Library*. Web. 10 July 2016.

Batllori M: *Borgia (Borja)*, 2nd ed, *New Catholic encyclopedia*, 2, Detroit, 2003b, Gale, *Gale Virtual Reference Library*. Web. 10 July 2016.

Burchard J: *Pope Alexander VI and his court—extracts from the Latin Diary of John Burchard*, Kindle Edition, Moorestown, NJ, 2015a, Perennial Press, Kindle Location 1959, (Kindle Locations 1959–1961).

Burchard J: *Pope Alexander VI and his court—extracts from the Latin Diary of John Burchard*, Kindle Edition, Moorestown, NJ, 2015b, Perennial Press, Kindle Location 1956–1957.

Burchard J: *Pope Alexander VI and his court—extracts from the Latin Diary of John Burchard*, Kindle Edition, Moorestown, NJ, 2015c, Perennial Press, Kindle Location 1861–1863.

Burford N, Carpenter Y, Conrad E, Saunders CD: The chemistry of arsenic, antimony and bismuth. In Sun H, editor: *Biological chemistry of arsenic, antimony and bismuth*, New York, NY, 2011, Wiley, p. 14.

Carroll James: *Constantine's sword: the church and the Jews—a history*, Boston, 2002, Houghton Mifflin, pp. 363–364.

Cloulas I. *The Borgias* (trans. G. Roberts) New York, NY: Franklin Watts; 1989a; p. 240.

Cloulas I. *The Borgias* (trans. G. Roberts) New York, NY: Franklin Watts; 1989b; p. 139.

Cloulas I. *The Borgias* (trans. G. Roberts) New York, NY: Franklin Watts; 1989c; p. 241.

Cloulas I. *The Borgias* (trans. G. Roberts) New York, NY: Franklin Watts; 1989d; p. 241–242.

Cloulas I. *The Borgias* (trans. G. Roberts) New York, NY: Franklin Watts; 1989e; p. 255–257.

Cobb C, The Chemistry of Lucrezia Borgia, et al: In Patterson GD, Rasmussen SC, editors: *Characters in chemistry: a celebration of the humanity of chemistry*, Oxford, UK, 2013a, Oxford University Press, p. 65.

Cobb C, The Chemistry of Lucrezia Borgia, et al: In Patterson GD, Rasmussen SC, editors: *Characters in chemistry: a celebration of the humanity of chemistry*, Oxford, UK, 2013b, Oxford University Press, p. 66.

Cobb C, The Chemistry of Lucrezia Borgia, et al: In Patterson GD, Rasmussen SC, editors: *Characters in chemistry: a celebration of the humanity of chemistry*, Oxford, UK, 2013c, Oxford University Press.

Cobb CL: *Magic, mayhem, and mavericks*, Amherst, NY, 2002, Prometheus Books.

Cobb CL, Fetterolf ML: *The joy of chemistry*, Amherst, NY, 2010, Prometheus Books, part II Chapter 5.

Cobb CL, Goldwhite HG *The chemistry of alchemy*, Amherst, NY: Prometheus Books; Chapter 8 and references therein.

De Dalmases C: *Borgia, Francis, St.*, 2nd ed, *New catholic encyclopedia*, 2, Detroit, 2003, Gale, *Gale Virtual Reference Library*. Web. 10 July 2016.

Eaton DL, Gilbert SG: In Casarett LJ, Doull JD, editors: *Toxicology: the basic science of poisons*, 8th ed, New York, NY, 2013, McGraw-Hill, pp. 13–49.

Hibbert C: *The Borgias and Their Enemies 1431–1519*, Boston, MA, 1994, Mariner Books.

http://historynewsnetwork.org/article/696.

https://www.uihealthcare.org/content.aspx?id=236459.

Lane RW: The Wissenschaften of toxicology: harming and helping through time. In Hayes AW, Kruger CL, editors: *Hayes' principles and methods of toxicology*, Boca Raton, FL, 2014, CRC Press, p. 14.

Lucas-Dubreton J. *The Borgias* (trans. PJ Stead) New York: NY: Dutton; 1954a; p. 240.

Lucas-Dubreton J. *The Borgias* (trans. PJ Stead) New York: NY: Dutton; 1954b; p. 175.

Lucas-Dubreton J. *The Borgias* (trans. PJ Stead) New York: NY: Dutton; 1954c; p. 263.

Lynch ME: Joan of Arc (c. 1412–1431). In Anne Commire, editor: *Women in world history: a biographical encyclopedia*, Vol. 8, Detroit, 2002, Yorkin Publications, pp. 185–193. Gale Virtual Reference Library. Web. 11 July 2016.

Mallett M: *The Borgias: the rise and fall of a Renaissance dynasty*, New York, NY, 1981a, Barnes and Noble.

Mallett M: The Borgias: the rise and fall of a Renaissance dynasty, New York, NY, 1981b, Barnes and Noble.

Mann J: Murder, magic, and medicine, Oxford, UK, 1992, Oxford University Press, p. 23–291, 51–54.

Mathew A: *The life and times of Rodrigo Borgia, Pope Alexander VI*, Kindle Edition, London, UK, 1912a, Endeavor Press.

Mathew A: *The life and times of Rodrigo Borgia, Pope Alexander VI*, Kindle Edition, London, UK, 1912b, Endeavor Press.

Mathew A: *The life and times of Rodrigo Borgia, Pope Alexander VI*, Kindle Edition, London, UK, 1912c, Endeavor Press.

Mathew A: *The life and times of Rodrigo Borgia, Pope Alexander VI*, Kindle Edition, London, UK, 1912d, Endeavor Press.

Mathew A: *The life and times of Rodrigo Borgia, Pope Alexander VI*, Kindle Edition, London, UK, 1912e, Endeavor Press.

Mathew A: *The life and times of Rodrigo Borgia, Pope Alexander VI*, Kindle Edition, London, UK, 1912f, Endeavor Press.

Mathew A: *The life and times of Rodrigo Borgia, Pope Alexander VI*, Kindle Edition, London, UK, 1912g, Endeavor Press.

Meyer GJ: *The Borgias: the hidden history*, Kindle Edition, New York, NY, 2013, Random House Publishing Group.

Meyer GJ: *The Borgias: the hidden history*, Kindle Edition, New York, NY, 2014a, Random House, Chapter 11.

Meyer GJ: *The Borgias: the hidden history*, Kindle Edition, New York, NY, 2014b, Random House, Kindle Location.

Meyer GJ: *The Borgias: the hidden history*, Kindle Edition, New York, NY, 2014c, Random House, Kindle Locations.

Noel G: *The Renaissance Popes: culture, power, and the making of the Borgia Myth*, Kindle Edition, Boston, MA, 2016a, Little, Brown Book Group, Kindle Locations.

Noel G: *The Renaissance Popes: culture, power, and the making of the Borgia Myth*, Kindle Edition, Boston, MA, 2016b, Little, Brown Book Group, Kindle Locations.

Norman NC: *Chemistry of arsenic, antimony and bismuth*, New York, NY, 1977, Springer, p. 191, 411.

O'Mara S: *Why torture doesn't work: the neuroscience of interrogation*, Cambridge, MA, 2015, Harvard University Press.

Oehme FW, Brown JF, Fowler ME: Toxins of animal origin. In Casarett LJ, Doull JD, editors: *Toxicology: the basic science of poisons*, New York, NY, 1975, Macmillan, pp. 580–588.

Partner Peter: Papal financial policy in the Renaissance and counter-reformation, *Past Present* 88:17–62, 1980. Web. http://www.jstor.org/stable/650552.

Patai R: *The Jewish alchemists: a history and source book*, Princeton, NJ, 1994a, Princeton University Press, p. 351.

Patai R: *The Jewish alchemists: a history and source book*, Princeton, NJ, 1994b, Princeton University Press, p. 208, 307, 211.

Scarsbrook MG: *The life & legend of Lucrezia Borgia*, Kindle Edition, Atlanta, 2011a, Red Herring.

Scarsbrook MG: *The life & legend of Cesare Borgia*, Kindle Edition, Seattle, WA, 2011b, Amazon, Kindle Locations.

Scarsbrook MG: *The life & legend of Cesare Borgia*, Kindle Edition, Seattle, WA, 2011c, Amazon, Kindle Locations.

Scarsbrook MG: *The life & legend of Cesare Borgia,* Kindle Edition, Seattle, WA, 2011d, Amazon, Kindle Locations.

Scarsbrook MG: *The life & legend of Cesare Borgia,* Kindle Edition, Seattle, WA, 2011e, Amazon, Kindle Locations.

Scott John, Sullivan Vickie: Patricide and the plot of the prince: Cesare Borgia and Machiavelli's Italy, *Am Polit Sci Rev* 88:899, 1994. Web. http://www.jstor.org/stable/2082714.

Stevens S: *Deadly doses, a writer's guide to poisons,* Cincinnati, OH, 1990a, Writer's Digest Books.

Stevens S: *Deadly doses, a writer's guide to poisons,* Cincinnati, OH, 1990b, Writer's Digest Books, p. 14, 17, 47.

Stevens S, Bannon A: *HowDunit: the book of poisons,* KindleEdition, Cincinnati, OH, 2007, Writers Digest Books.

Thomas H: The slave trade: the story of the Atlantic slave trade: 1440–1870, New York, NY, 1999, Simon & Schuster.

Thompson CJS: *Poisons and poisoners,* New York, NY, 1993, Barnes and Noble, Chapters 8.

von Hohenheim P: In Casarett LJ, Doull JD, editors: Toxicology: the basic science of poisons, New York, NY, 1975, Macmillan. Opening quote.

# CHAPTER 6

# Aqua Tofana

**Mike Dash**
University of Cambridge, Cambridge, United Kingdom

Aqua Tofana was the name given to a poison that, contemporary accounts suggest, was first created in Sicily in about 1630 and was widely used in Rome in the middle of the 17th century. The poison was made and dispensed by a group of "wise women" to an almost exclusively female clientele and was employed primarily to murder cruel or unwanted husbands. The surviving members of this group were rounded up and tried in Rome in 1659.

The historical Aqua Tofana appears to have been a solution with a base of arsenic, antimony, and lead (Ademollo, 1881), possibly with the addition of "corrosive sublimate," the contemporary term for mercuric chloride (Gigli, 1958). It is impossible, in a period in which sudden death was common and forensic science in its infancy, to determine how widely the poison was used, and how many of the deaths attributed to it were actually the product of natural causes. However, the notoriety of the poison was such that, by the early 18th century, "Aqua Tofana" had become a catch-all term used to describe a supposed class of subtle, precise, slow-acting poisons that were believed to be undetectable and invariably fatal. There was widespread belief in the existence of such "slow poisons" in the 18th and early 19th centuries, and rumors of their use were common in cases of sudden or unexpected death; the dying Mozart stated his belief that he had been poisoned with Aqua Tofana (Karhausen, 2011). There is, however, no evidence that slow poisons actually existed (Dash, 2015).

## 6.1 HISTORY

The first recorded mention of Aqua Tofana dates to 1632–33, when two poisoning trials took place in Palermo, Sicily. In the first, which took

Toxicology in the Middle Ages and Renaissance. DOI: http://dx.doi.org/10.1016/B978-0-12-809554-6.00006-8

place in February 1632, a woman named Francesca la Sarda was executed for using a poison that killed its victim in 3 days. In July 1633, a second woman, Teofania di Adamo, was put to death for a similar crime.

These cases were reinvestigated in 1881 by a Sicilian antiquary, Salvatore Salomene-Marino, who uncovered contemporary chronicles and diaries that outlined the two cases. According to Salomene-Marino, La Sarda and Di Adamo had worked together to make and sell a poison known as "Acqua Tufània." He hypothesized that Di Adamo was the inventor of the poison, which was named after her, and that La Sarda had been her assistant (Salomene-Marino, 1881). The extreme punishment meted out to Di Adamo suggests that her crime was considered an especially cruel and vicious one; competing accounts state that she was either hanged, drawn, and quartered (Salomene-Marino, 1881) or "closed and bound, alive, in a canvas sack... [and] thrown from the roofs of the bishop's palace into the street in the presence of the populace" (Boccone, 1697).

Other associates of Di Adamo and La Sarda who survived the Palermo round-up subsequently fled to Rome, where they continued to manufacture and sell Aqua Tofana (Ademollo, 1881). This group was led by Giulia Tofana, who may have been the daughter of Teofania di Adamo. She was accompanied by a much younger woman named Girolama Spara and recruited other accomplices in Rome who had a knowledge of the city and its people. According to an investigation that took place in the late 1650s, Tofana's gang obtained a supply of arsenic via a priest, Father Girolamo of Sant'Agnese in Agone, a church in the center of Rome. Father Girolamo's brother was an apothecary with access to the poison (Salomene-Marino, 1881; Ademollo, 1881).

Tofana died in about 1651—probably in her own bed, and apparently unsuspected of any crime—and Spara took over as leader of the gang. She was the widow of a Florentine gentleman and moved comfortably in aristocratic circles, while an associate, Giovanna de Grandis, dealt with clients from the lower social classes. A later investigation suggested that the two women took the arsenic supplied by Father Girolamo and disguised it by turning it into a liquid and bottling it in glass jars that identified it as "Manna of St Nicholas"—then a popular healing oil collected from the spot where it was said to drip from the saint's bones in a church in Bari. The preparation was sold under the cover of it being a preparation that removed blemishes from faces (Ademollo, 1881).

The gang now led by Spara finally came to the attention of the Roman authorities in 1658. A contemporary *Life* of the then Pope, Alexander VII, suggests that word of the poison first leaked out in the confessional (Sforza-Pallavicino, 1838–45); court records privilege a version of events in which de Grandis was entrapped by a police agent who told tales of an unhappy marriage and offered large sums for a poison that would kill her husband. The trial of the remaining members of the gang heard evidence of 46 murders. Five ringleaders, including Spara and de Grandis, were hanged in front of an unusually large crowd in July 1659; one of their clients was also executed (Ademollo, 1881). Six accomplices and more than 40 of the gang's lower class customers were tried at the same time, most of whom were imprisoned for life (Sforza-Pallavicino, 1838–45). It would appear that both Father Girolamo and the gang's aristocratic clients escaped these punishments. The historian Alessandro Ademollo uncovered evidence suggesting that the names of several more socially prominent victims were deliberately kept out of the trial on the order of the Pope. The best known of these was the Duke of Ceri, a well-connected Roman nobleman who was said to have been poisoned by his much younger wife. The Duchess escaped punishment, but was ordered to remarry (Ademollo, 1881).

## 6.2 SYMPTOMS

The symptoms produced by Aqua Tofana were set out in a public notice issued in Rome at the time of the trial. They included burning pain in the throat and stomach, vomiting, extreme thirst, and diarrhea, all of which point to arsenic as the active ingredient of the poison. The suggested antidote was lemon juice and vinegar.

Aqua Tofana was described as clear and tasteless, suggesting that a key part of the manufacturing process was masking the characteristic metallic taste of arsenic. It was also considered to be a relatively "gentle" poison, which did not produce so much vomiting as—and hence aroused less suspicion than—other preparations known at the time (Ademollo, 1881).

## 6.3 POISONS AND THE CRIMINAL MAGICAL UNDERWORLD

The details brought out in the trial of 1659 make it possible to reconstruct the ways in which preparations such as Aqua Tofana were sought and used in an early modern European capital like Rome. The manufacture and sale of poisons was only one small part of the activities of an extensive underworld that specialized in the supply of services that could not be offered by the established church and state. This community may have been at least 200 strong in mid-17th century Rome, and numbered among its members wise women, astrologers, alchemists, confidence men, witches, shady apothecaries, and back-street abortionists who between them told fortunes and cast horoscopes, sold love potions and lucky charms, cured toothache, and offered to dispose of unwanted babies and unwanted husbands (Dash, 2015).

The historian Lynn Wood Mollenaeur, who wrote on the existence of a similar community in Paris a quarter of a century later, coined the term "criminal magical underworld" to describe it. In the deeply religious societies of the time, such underworlds centered on renegade priests who were prepared to use religious rituals—up to and including the so-called "amatory" or black mass—to harness sacerdotal power and create what they and their clients believed to be effective potions and spells. Mollenaeur identifies five such renegades in Paris, among them a 70-year-old priest, Abbé Étienne Guibourg, who led an active double life as a black magician and blessed the ingredients, intended for love philters and other similar products sold by the self-proclaimed "magicians," "wise men," and "cunning women" who made up the bulk of the local magical underworld. Raids conducted by the police on the Paris under-world of the 1680s uncovered quantities of grimoires, incense, wands, and a rich variety of ingredients used in sexual magic, including breast milk and bags of powdered menstrual blood (Mollenaeur, 2006).

Father Girolamo, the priest who sourced arsenic for Giulia Tofana's poisoning gang, seems to have fulfilled some of these same functions in Rome (Dash, 2015), but other such renegades also existed (Rietbergen, 2006). The operation led by Tofana and Spara offered many services other than Aqua Tofana, and Spara also operated as a "cunning woman" who sold charms and cures to the gentlewomen and nobility of Rome.

Women of this kind not only drew on a common magical tradition that had existed in Europe since the Middle Ages, but also possessed the means to attract clients; new customers were typically drawn into the magical underworld by way of a minor transgression involving a consultation with a wise woman—perhaps seeking the recovery of some lost valuable or the fun of having a fortune told. They might then be induced to progress to more costly services such as fertility treatments or abortion. All these activities would not only have introduced Tofana and Spara to women who might become regular customers but would also have given them a shrewd idea of which of their clients were happy in their marriages, which unhappy—and which desperate enough to seek drastic remedies. The everyday activities of a criminal magical underworld thus helped to funnel customers willing to pay high prices for poisons to the people able to supply such products to them (Mollenaeur, 2006; Dash, 2015).

## 6.4 THE SLOW POISONS

The notoriety generated by the trial of 1659 was such that "Aqua Tofana" and "water from Palermo" became generic terms for an especially deadly poison. Both terms can be found in a wide range of sources, from medical textbooks to court records, for more than 200 years. The case also reinforced existing prejudices, most notably the ideas that that poison was a quintessentially female weapon and that Italians were more skilled than the people of any other nation in its use (Fiume, 2009).

Reports suggesting that the secret of making Aqua Tofana had survived the deaths of Tofana, Spara and their gang were commonplace throughout the 18th century. The French traveler Jean-Baptiste Labat described the capture and execution of an old woman who sold bottles of a clear poison disguised as saint's manna in Naples in 1709 (Labat, 1730). Pius Nikolaus von Garelli, who was personal physician to the Holy Roman Emperor Charles VI (reigned 1711–40), reported reading legal papers in the possession of his master that attributed the deaths of 600 people to the use of Aqua Tofana at around the same time (Hoffman, 1718–34). And Johann Keysler, a Fellow of the Royal Society, wrote in 1730 of an elderly female poisoner known by the name "Tophana" who had murdered several hundred people using the same liquid and was held in prison in Naples (Keysler, 1778).

The same group of accounts, notably Keysler's, also played an important part in elevating the reputation of Aqua Tofana and turning it into something it had clearly not been in the first half of the previous century. In this new conception, Aqua Tofana was reimagined as an especially precise and insidious poison in a process that can be traced across the period 1700–1850.

In its most developed form, the process resulted in claims for a poison that was very widely feared, but never actually existed. This Aqua Tofana was colorless, odorless, and tasteless, and hence undetectable; it was lethal in extremely small doses; and it was possible, by controlling the dose, to ensure that it killed within a prescribed period of time, which might be anything between 3 days and several months. It was also so subtle that death would generally be ascribed to natural causes. Collectively, these characteristics came to define a class of toxicants known as "slow poisons," of which Aqua Tofana was by far the best known and, supposedly, the most dangerous. The 19th century writer Johann Beckmann defined this group as "all those preparations which can be administered imperceptibly, and which gradually weaken the vital powers, and finally cut short the life of a man" (Beckmann, 1823).

An especially complete description of Aqua Tofana as a slow poison appeared in *Chambers's Journal* in 1890:

> *Administered in wine or tea or some other liquid by the flattering traitress, [it] produced but a scarcely noticeable effect; the husband became a little out of sorts, felt weak and languid, so little indisposed that he would scarcely call in a medical man....*
>
> *After the second dose of poison, this weakness and languor became more pronounced... The beautiful Medea who expressed so much anxiety for her husband's indisposition would scarcely be an object of suspicion, and perhaps would prepare her husband's food, as prescribed by the doctor, with her own fair hands. In this way the third drop would be administered, and would prostrate even the most vigorous man. The doctor would be completely puzzled to see that the apparently simple ailment did not surrender to his drugs, and while he would be still in the dark as to its nature, other doses would be given, until at length death would claim the victim for its own...*
>
> *To save her fair fame, the wife would demand a post-mortem examination. Result, nothing—except that the woman was able to pose as a slandered innocent, and then it would be remembered that her husband died without either pain, inflammation, fever, or spasms. If, after this, the woman within a year or two formed a new connection, nobody could blame her; for, everything considered, it would be a sore trial for her to continue to bear the name of a man whose relatives had accused her of poisoning him. (Anon, 1890)*

Aqua Tofana thus completed a transition from an apparently real poison, based on an identifiable base, that produced relatively violent symptoms and was comparatively readily detected, to a nonexistent super-poison that was far more widely feared and cited—one that "the acutest analysts were utterly unable to testify to its presence in the organs of one of its victims after the most searching post-mortem examination. It was, in fact, the poisoner's *beau-idéal* of a poison" (Anon, 1890). Barely remembered today, its origins have been almost entirely obscured.

## REFERENCES

Ademollo A: *I Misteri dell'Acqua Tofana*. Rome, 1881, Tipografia dell'Opinione.

Anon: Tofana, the Italian poisoner, *Chambers's J*, 1890.

Beckmann J: *A concise history of ancient institutions, inventions and discoveries*, London, 1823, G & W.B. Whittacker.

Boccone P: Museo di Fiscia e di Esperienze Variato, e Decorato di Osservazioni Naturali, Venice, 1697, Baptistam Zuccato.

Dash M: Aqua Tofana: slow-poisoning and husband-killing in 17th century Italy, mikedashhistory.com, accessed October 31, 2015.

Fiume G: The old vinegar lady. In Laveck B, editor: *New perspectives on witchcraft*, London, 2009, Routledge, pp. 261–286.

Gigli G: *Diario Romano, 1608–1670*, Rome, 1958, Editore Tuminelli.

Hoffmann F: *Medicinæ Rationalis Systematicæ*, Halle, 1718–34, Officina Rengeriana.

Karhausen L: *The bleeding of Mozart*, Bloomington, 2011, Xlibris Corporation.

Keysler G: *Travels through Germany, Hungary, Bohemia, Switzerland, Italy and Lorraine*, London, 1778, J. Scott.

Labat J-B: *Voyages du P.[ere] Labat de L'Ordre des FF. Prescheurs, en Espagne et en Italie*, Paris, 1730, JP & CJP Delespine.

Mollenaeur LW: *Strange revelations: magic, poison, and sacrilege in Louis XIV's France*, University Park, PA, 2006, University of Pennsylvania Press.

Rietbergen PJAN: *Power and religion in Baroque Rome*, Leiden, 2006, Brill.

Salomene-Marino S: L'Acqua Tofana, *Nuove Effemeridi Siciliane* 11:285–294, 1881.

Sforza-Pallavicino P: *Vita Di Alessandro VII*, Milan, 1838–45, Giovanni Silvestri.

# CHAPTER 7

# Poisons and the Prince: Toxicology and Statecraft at the Medici Grand Ducal Court

Sheila Barker
The Medici Archive Project, Florence, Italy

## 7.1 POISONS IN 16TH-CENTURY SOCIETY

During the centuries that the Medici family reigned over Florence and then Tuscany (1531−1743), poison was a major preoccupation of Europe's ruling elite. Letters sent to Grand Duke Cosimo I de' Medici (1518−74) (Fig. 7.1) speak continually of poisoning as an assassination method. Citing symptoms such as red rashes, bleeding from all the orifices of the head, vomiting of food and blood, blackened skin, and extreme postmortem edema, Cosimo's correspondence implicates poison in the deaths of Charles IX of France, Cardinal Charles de Lorraine, Countess Bianca Ragnoni Guidi, and Cardinal Ippolito de' Medici; it blames poison for the illnesses of Queen Mary of Scotland, Cardinal Charles Borromeo, and the monks in a monastery in Sansepolcro; and it warns of the plots to poison Queen Elizabeth I, King James I, Prince Andrea Doria, Marchioness Pentelisea Dal Monte Santa Maria, King Philip II of Spain, and his consort Elisabeth de Valois. Pope Paul III and Henry II of France were said to be attempting to poison Ferrante Gonzaga, who in turn was conspiring with Cosimo to taint the wine flask of their common enemy, Piero Strozzi. Cosimo himself was the target of poison plots: In 1556, a letter alerted him to the danger of poison-laden hand towels, and 3 years later another letter warned him that a trusted woman was trying to render his food and drink deadly.

Contributing to the frequency of poisonings were the many means of carrying them out. In Renaissance Tuscany these included venoms from snakes, scorpions, toads, and fish; phytotoxins from hemlock, black henbane, oleander, aconite, poppy, nox vomica, black nightshade, and white hellebore; and mineral and alchemical poisons such as arsenic, gypsum, and the mercury salt called *solimato*. In addition,

Toxicology in the Middle Ages and Renaissance. DOI: http://dx.doi.org/10.1016/B978-0-12-809554-6.00007-X

*Figure 7.1 Lodovico Cardi (Il Cigoli). Portrait of Grand Duke Cosimo de' Medici. 1602–03. Oil on canvas, 395×215 cm. Palazzo Medici-Riccardi Prefettura, Florence, Italy. Photo by Kaho Mitsuki, Image in the Public Domain (PD-1996).*

poisonings could occur in the absence of a human agent, whether through the bite of a rabid dog or a breath of pestilential air. Poisonous air, moreover, was a danger not only in times of plague, but also anytime that a rotten smell could be detected since, according to Renaissance medical theory, the unpleasant odors released by decaying, corrupt matter were themselves agents of morbid putrefaction. During Cosimo's reign, Tuscany experienced several public health crises due to poisonous air. In one case, putrid smells issuing from Pistoia's Hospital of the Ceppo (Fig. 7.2) began sickening the nuns in the monastery next to it; in another case, Cosimo's dredging of the harbor at Livorno in 1546 released a terrible stench that local authorities blamed for the port

*Figure 7.2 Hospital of the Ceppo (Ospedale del Ceppo). Pistoia, Italy. Photo by Jollyroger, licensed through Creative Commons CC-BY-SA-3.0.*

city's subsequent surge in death rates and decline in birth rates; and, in a parallel situation, when Cosimo attempted to reclaim the swamplands known as the Padule di Fucecchio by planting an orchard, the local peasants protested that his fruit trees were rotting and exuding pestilential fumes that in turn were the cause of a genocidal die-off in the surrounding villages.

As plentiful as the poisons were the individuals who trafficked in them. Physicians and apothecaries used them frequently, and were expected to take precautions such as storing potential poisons in locked containers, and adjusting dosages according to the patient's physical condition, such as with the sleeping drug made from nightshade whose recipe warns that "this medicine is very dangerous, and should only be used on those in very robust health" (Ms, Marc. It. III.10). Another category of learned professionals with expertise in poisons, particularly phytotoxins, were the herbalists whom the grand dukes employed at their two botanical gardens and at their university in Pisa. Renowned for having such knowledge was the duke of Alba's herbalist. He specialized in making poisons for hunting large game, four powder kegs of which were sent as a gift to Cosimo's son, Francesco de' Medici (1541–87). Like hunting, warfare also entailed the use of toxins. Soldiers coated their knives and the shot of their harquebuses (an early style of long gun) with poison in order to compound the lethality of their weapons, and it was a common military strategy to poison the wells in enemy territory. Suppliers to a wide cross-section of society were the shadowy people whose knowledge of toxicology was more casually acquired. Examples are the woman known as Girolama Venefica and the sorcerer named Apollonio, societal outliers who ran a grave risk of being scapegoated whenever murder by poison was committed in their vicinity.

Even the ordinary subjects of the Medici grand duchy were equipped to concoct poisons. Artisans and artists worked on a regularly basis with toxins such as the poison sumac used to make wood varnish, the mercury used to refine gold, and the cinnabar, lead, and minium (a form of lead tetroxide) used in painting. Monastics often cultivated dangerous plants in their cloister gardens and, still etched in communal memory, was the occasion when the nuns of Santa Petronilla in Perugia had poisoned Pope Benedict XI in 1304 by means of laced figs (Ms, BNCF, Magl. XXV, 18). Even children knew about poisons, as demonstrated by the case of a girl in Sienna who, fearing that she would be forced into a nunnery, fed her family a salad sprinkled with deadly mercury salts and quicksilver that her mother had obtained from a milliner for cosmetic purposes and which she had kept in a lockbox.

## 7.2 KNOWLEDGE OF POISONS AT THE MEDICI COURT

Several generations of the Medici family possessed an empirical knowledge of poisons. In the 15th century, these included banking tycoon and alchemical tinkerer, Cosimo "Pater Patriae" de' Medici (1389–1464) and Grand Duke Cosimo de' Medici's paternal grandmother Caterina Sforza (1463–1509), whose prodigious collection of alchemical, medical, veterinary, and magical recipes was handed down through the family. Although Grand Duke Cosimo de' Medici sent a letter in 1548 in which he foreswore any desire to abet a scheme to poison someone ("things that horrify us") and protested that he had no relevant recipes, an internal court memo written immediately afterwards proves that the duke's letter was deliberately composed to deflect suspicion in case of interception. Cosimo was not, in fact, above suspicion: In 1566, malicious gossip accused him of poisoning his chamberlain, Sforza Almeni.

In truth, Cosimo not only knew of the poison-brewing sorcerer named Apollonio, but he also kept a poison recipe among his important confidential documents. Furthermore, he owned several books and manuscripts in which poisons were discussed, including Dioscorides'

*Materia Medica,* Pliny's *De Natura Rerum,* various Galenic writings, a well-annotated copy of Pietro Andrea Mattioli's *Discorsi,* a Latin manuscript roughly based on Pietro d'Abano's *Liber de Venenis,* and a vulgate abridgement of Abano's treatise entitled "Trata de li veneni." Recipes for poisonous compositions also crop up in the grand duke's alchemical treatises and recipe collections, such as the formula for a "perfect temper for armor" that warns the alchemist to "watch out not to cut yourself (while handling the temper) in any way, because it will be incurable as this is extremely poisonous" (Ms, BNCF, Strozz. XIX, 94). In the margin next to this formula, Cosimo wrote "veleno" ("poison"), indicating perhaps what he considered to be a more useful application of the temper recipe.

Cosimo's children pursued their father's interest in poisons. The elder son, Francesco, noted above for having received hunting poison as a gift, was so well versed in toxicology that when one of his courtiers, Count Clemente Pietra, was murdered in 1574, Francesco deduced that the knife which had wounded him must have been coated with a Spanish poison. Francesco was also the dedicatee of Bacci's (1573) treatise on the unicorn horn as an antidote (a subsequent edition was dedicated to Francesco's second wife). Acceding to the throne after Francesco's untimely death—one which has always been overshadowed by suspicions of poisoning—Cosimo's younger son Ferdinando (1549–1609) gave patronage to a physician with unsurpassed toxicological expertise: Girolamo Mercuriale, the author of the 1584 *De venenis, et morbis venenosis tractatus locupletissimi (Mercuriale, 1584).* Ferdinando's personal and pragmatic toxicological knowledge is revealed in a letter he sent, along with some poison, to an agent in 1590. Ferdinando's letter explains how to slip the poison into the enemy's wine flask, and touts the composition as being odorless, tasteless, and sufficiently potent to kill with just one tainted glass of wine. Francesco's orphaned son, Don Antonio de' Medici (1576–1621), dedicated himself to alchemy and his recipes included a formula for "the herb for the crossbow," made from the root of white hellebore (Ms, BNCF, Naz. II.I.345).

## 7.3 MEDICI ANTIDOTES

The fear of poisons sometimes induced paranoia and ferocious intolerance among the ruling elite. Thus, it was not an isolated case when the Italian military commander Alfonso d'Avalos d'Aquino, Marquis of

*Figure 7.3 Benvenuto Cellini,* Cellini Salt Cellar *(also known as the* Saliera*). 1543. Partially enameled gold. Kunsthistorisches Museum, Vienna, Austria. Photo by Vassil, licensed through Creative Commons CC0-1.0.*

Pescara, had a servant drawn and quartered in 1546 merely for being in possession of a poison. The Medici, however, made recourse to laws rather than violence in order to combat poisoning. Notably, Duke Alessandro de' Medici issued a decree in 1531 prohibiting Florentine apothecaries from supplying anyone, regardless of rank or station, with arsenic, *solimato*, realgar, or any other poison, and mandating that poisons be kept locked away. The future Medici grand dukes extended Alessandro's laws throughout all of Tuscany, where fines or jail—not death—remained the punishment for unlawful possession of poison (Ms, BNCF, Targioni Tozzetti 189, cod. 5).

Besides legislating controlled substances, the Medici also put a phalanx of safeguards into place around their banquet tables. They chose their kitchen staff carefully, employing a *scalco* and a *bottigliere* to ensure the safety of the food and wine. Out of scruple, the grand duchesses maintained their own kitchen staff, recruiting from their homelands to foster these servants' loyalty. As an additional precaution, some of the vessels on the dinner table served for the detection and neutralization of toxins. Open-air salt cellars with shiny silver and gold surfaces, such as the magnificent *Cellini Salt Cellar* (Fig. 7.3) created by Florentine goldsmith Benvenuto Cellini in 1543, were just such mechanisms. Their utility is explained in the treatise *De venenis* by physician

Antonio Guainerio, who asserted that air "sweats" in the presence of a poison, especially in the vicinity of salt, pointing out that a shiny metallic surface allows one to see this sweat; moreover, according to Guainerio, the salt itself counters the effects of poison (Thorndike, 1934).

Another line of defense was provided by the opulent goblets from which the Medici drank, since many of these were wrought from alexipharmic stones. A prime example is the kylix (a wide and shallow cup with horizontal handles) given by Catherine de' Medici to her granddaughter Christine de Lorraine, who then brought the cup to the Medici court as part of her wedding trousseau (Magnificenza alla corte dei Medici, 1997). It was made of the green chalcedony called plasma, which loses its luster upon contact with a poison according to Mercati's (1576) treatise, *Istruttione sopra i veleni*. The cup that Cosimo I de' Medici received from Mughal Emperor Humayun at the hands of a Marrano jewel merchant named Alvaro Mendes was made of rhinoceros horn, a material regarded as powerful antidote when even a tiny amount of its filings were taken with wine (Bacci, 1573; Mercati, 1576). Since the true animal origins of purported rhinoceros and unicorn horns were difficult to assess, distinguished provenances were once the best argument for these objects' authenticity and efficacy. Thus, in 1546, Cosimo de' Medici decided to purchase a unicorn horn when he learned that it had been previously owned by two of his ancestors, Pope Clement VII de' Medici and Cardinal Ippolito de' Medici.

Aside from such rarities, the majority of the Medici's precious alexipharmic vessels were made either from agate—a cure for scorpion and viper venom, according to Bacci's (1587) treatise, *Le XII pietre pretiose*—or from lapis lazuli. Notably, the latter was not utilized for vessels at the Medici court until the late 16th century. The reason for the Medici's diffidence may have been lingering associations of lapis lazuli with poison. Although Galen had endorsed this stone, Avicenna and Averroes had denigrated it as a corrosive poison (Mattioli, 1548, "Della Pietra Cerulea"). Similarly, the two above-mentioned treatises derived from Pietro d'Abano's writings had designated lapis lazuli in its untreated form as a toxicant that injured the stomach and caused symptoms of acute melancholy. The subsequent reintegration of lapis lazuli among *materia medica* was probably instigated by Mattioli's argument in his 1548 *Discorsi* that Arabic physicians had been confused about the stone's identity. Sanctioned by Mattioli, treated lapis lazuli was included in Florence's official pharmacopeia of 1560 (Ricettario, 1560).

Of all the Medici dynasty's measures against poisoning, the most consequential was their production of the antidote known as "Olio contro a veleni del Granduca." Both a prophylactic and a cure, its distinguishing ingredient was a laboriously prepared oil made from scorpion venom (although its opium content was not to be discounted). Grand Duke Cosimo I was hailed as its inventor. However, its formula resembles earlier antidotes, including one attributed to Mesue, one from Mattioli's *Discorsi*, and one named "Oleum Clementis" after Pope Clement VII de' Medici, who had distributed it during the plague of 1527 (Falloppio, 1606; see also Perifano, 1997).

Grand Duke Cosimo I never sold his antidote; rather, whenever allies and friends needed it, Cosimo sent the drug by courier, packaged in glass vials that were tucked into the compartments of customized wooden cases. Soon it was requested by nobles throughout Europe, including the duke of Bivona, the duke of Montmorency, the duchess of Alba, and the count of Frankenburg, who thanked the grand duke with gifts of rare stones. It did not hurt the drug's reputation that Mercati, in his 1576 *Lesson on Poisons*, classified the "oil made by the Grand Duke of Tuscany" as an antidote that "resolves or evaporates the poison." By the late 16th century, Medici ambassadors stocked it at their outposts as a ready diplomatic gift for foreign rulers beset by plague, incurable fevers, poison, and, on occasion, any life-threatening condition for which no cure was known. Though the Medici grand duchy was a mere satellite state in the grand scheme of European politics, the priceless drug gave them an unexpected advantage in an age when it was a truism that "remedies against poisons should always be kept on hand by great princes" (Mercati, 1576).

## 7.4 TESTING POISONS

With such a valuable diplomatic tool in their hands, the Medici grand dukes were understandably concerned about its efficacy. As already noted, they packaged the scorpion-oil antidote carefully, both to resist breakage and to prevent degradation of the drug due to interaction with its container or temperature changes. Sometimes, the Medici grand dukes warned recipients that the oil was no longer fresh (being prepared only once a year in accord with the annual "maturation" of scorpion venom around August, when the dog star Sirius reappears in the night sky). Though there is no firm proof, it is possible that the

Medici tested the Olio del Granduca to ascertain its medicinal value. The testing of antidotes was not unprecedented in Renaissance Italy. Fallopio asserted that his teacher, Antonio Musa Brasavola, had tested the Oleum Clementis on a condemned criminal at the duke of Ferrara's command, and Andrea Bacci recounted how Cardinal Cristoforo Madruzzo had verified a unicorn horn's efficacy by administering some filings to one of two doves that had been fed arsenic (Falloppio, 1606; Bacci, 1573). Indeed, the idea of testing may have occurred to Cosimo in 1546 when a Tuscan peasant claimed to have found a unicorn horn growing inside a tree, and proved its authenticity by testing its antidotal powers on a dog that had been fed a poison. Two years later, in 1548, Cosimo sent his poison antidote to Duke Ferrante I Gonzaga and recommended that it be tried first with a prisoner awaiting capital punishment; aware that different batches might behave differently, he sent both old and new preparations of the oil as well as dosing instructions. The Medici grand duke himself called for such an experiment in 1566 in order to test a different antidote, Mattioli's "powder against all poisons" (based on the recipe in his *Discorsi*, bk. 6), a batch of which had been prepared a month in advance by the court apothecary, Geremia Foresti. After a convict on death-row agreed to take a mortal dose of arsenic followed by the experimental powder, the subject endured several days of excruciating intestinal agony while a panel of six physicians working in continual shifts carefully recorded his vital signs and prepared a medical report (Ms, ASF, Carte Strozz. I, 97; see also Marinozzi et al., 2015). By 1576 the Olio del Granduca itself was being tested at the Medici court. According to a Venetian ambassador's report from that year, Grand Duke Francesco de' Medici ordered prisoners awaiting their death sentences to drink poison, "and then using this oil he cured them" (Relazioni degli Ambasciatori veneti al Senato, 1916).

By the 17th century, the testing of potential new antidotes as well as the poisons themselves had captured the fascination of the Medici court. Girolamo Mercuriale recounted such an experiment performed under Grand Duke Ferdinando I on a man condemned to the gallows in Pisa, as a means of determining the efficacy of unicorn horn that someone wished to sell him; the poison, crystalline arsenic, was apparently effective, but the antidote failed (Mercuriale, 1589). In 1637, an elaborate poison experiment was carried out in Florence as "a noble and virtuous entertainment" to cheer the convalescent grand duke,

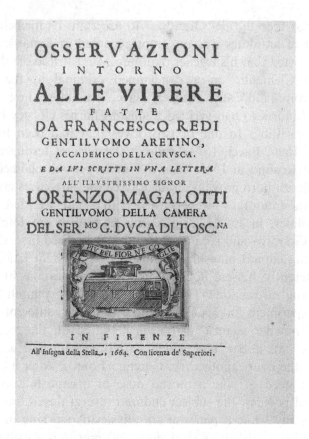

*Figure 7.4 Frontispiece: Francesco Redi,* Osservazioni intorno alle vipere. *Florence: All'Insegna della Stella, 1664. Image in the Public Domain (PD-1923).*

Ferdinand II. Turtles, geese, peacocks, goats and foxes were all bitten by the same viper, and after all the animals died as a result, the autopsies revealed blood clots in the heart and the veins. Following an impromptu debate over whether the clotting was the cause of death or merely a corollary symptom, several tests were done with vipers to determine where the venomous organ was located. During the course of these tests, the viper handler voluntarily drank all the venom that could be milked from the snake, he consumed its pulverized fangs, and he ate its liver as well, to show that these did not cause harm; but then he put a single drop of the same venom on a small cut made on a chicken, and the bird died after three hours, revealing that the venom's action was on the blood (Targioni Tozzetti, 1780). Some of these same experiments were repeated in Florence two decades later by Medici court physician Francesco Redi, only now under the rigorous testing methodology promoted by the Medici's Accademia del Cimento, of which Redi was a founding member. Redi's descriptions of these experiments, considered a milestone in experimental toxinology, were published first in 1664 and again in 1687 as *Osservazioni intorno alle vipere* (Redi, 1664) (Fig. 7.4). By this time, toxicology had become an openly discussed science and a point of pride for the Medici grand dukes. At the same time, poison itself managed to acquire an occasional benevolent face, providing aid in certain dire situations. For example, at the Medici grand dukes' hospital of Santa Maria Nuova, euthanasia by poison was carried out in 1654 and again in 1659 in cases of rabies, which was then incurable as well as agonizing for the patient and his or her caretakers. Thoroughly domesticated and exploited to every possible advantage, poisons had at last acquired a degree of scientific cachet under Medici tutelage.

## 7.5 RESOURCES

### 7.5.1 Manuscripts
Archivio di Stato di Firenze (ASF):

Santa Maria Nuova 149
Carte Strozziane, ser. I, filza 97

Biblioteca Nazional Centrale di Firenze (BNCF):

Magliabechiana XXV, cod. 458
Nazionale II.III.299
Palatina 548 "Trata de li veneni"
Strozziani XIX, cod. 94
Targioni Tozzetti 189, "Selva di Giovanni Targioni Tozzetti," cod. 5

Biblioteca Nazionale Marciana (Marc.): It.III.10

### 7.5.2 Databases
"BIA" published by the Medici Archive Project: www.bia.medici.org

## REFERENCES

Bacci A: *L'alicorno*, Florence, 1573, Giorgio Marescotti.

Bacci A: *De venenis et antidotis prolegomena seu communia praecepta ad humanam vitam tuendam saluuberrima in quibus diffinitiva methodus veneorum proponitur per genera, ac differentias suas partes, & passiones, praeservandi modum, & communia ad eorum Curationem Antidota complectens. De canis rabiosi morsu, et eius curatione*, Rome, 1586, Vincenzo Accolti.

Bacci A: *Le XII pietre pretiose, le quali per ordine di Dio nella Santa Legge, adornano I vestimenti del sommo Sacerdote. Aggiuntevi il diamante, le margarite, e l'oro poste da S. Giovanni nell'Apocalisse in figura della celeste Gierusalemme: con un sommario dell'altre pietre pretiose*, Rome, 1587, Bartolomeo Grassi.

Falloppio G: *Opera Genuina omnia, tam practica, quam theorica*, Venice, 1606, De Franciscis.

*Magnificenza alla Corte dei Medici*, 1997, Electa; Milan.

Marinozzi S, Giuffra V, Kieffer F: Baccio Baldini (1517–1589), protomedico alla corte medicea tra umanesimo e sperimentalismo, *Acta Med Hist Adriat* 13:345–364, 2015.

Mattioli PA: Discorsi *[* = Di Pedacio Dioscoride Anazarbeo Libri cinque Della historia, et materia medicinale tradotti in lingua volgare italiana da M. Pietro Andrea Matthiolo Sanese Medico, con amplissimi discorsi, et comenti, et dottissime annotationi, et censure del medesimo interprete*]*, 2nd ed., Venice, 1548, Vincenzo Valgrisi, Books 5 & 6.

Mercati M: Istruttione sopra i veleni *[* = Istruttione sopra la peste di Michele Mercati...aggiuntevi tre altre instruttioni: Sopra i veleni occultamente ministrati, Podagra & Paralisi*]*, Rome, 1576, Vincentio Accolto.

Mercuriale G: *De venenis, et morbis venenosis tractatus locupletissimi*, Venice, 1584, Paolo Meieto.

Mercuriale G: Consultationes *[* = Reponsorum et consultationum medicinalium in duo volumina digesta*]*, Vol. II, Venice, 1589, Iolitos.

Perifano A: *L'Alchimie à la cour de Côme I^er de Medicis: savoirs, culture et politique*, Paris, 1997, Honoré Champion.

Redi F: *Osservazioni intorno alle vipere*, Florence, 1664, All'Insegna della Stella.

*Relazioni degli ambasciatori veneti al Senato*, Vol. IV, Bari,1916, Laterza.

*Ricettario utilissimo et molto necessario a tutti gli spetiali, che vogliono preparar le medicine regolatamente, da diversi et eccellenti medici riveduto & approvato*, 1560, Vincenzo Valgrisi; Venice.

Targioni Tozzetti G: *Notizie degli aggrandamenti delle scienze fisiche accaduti in Toscana nel corso di anni LX. del secolo XVII*, Florence, 1780, Giuseppe Bouchard.

Thorndike Lynn: *A history of magic and experimental sciences*, Vol. IV, New York, 1934, Columbia University Press.

# CHAPTER 8

# Georgius Agricola, a Pioneer in the Toxic Hazards of Mining, and His Influence

Sverre Langård

Oslo University Hospital, Oslo, Norway

## 8.1 EDUCATION AND EARLY LIFE

Georg *Bauer* was born at Glauchau in the kingdom of Saxony, currently part of Germany, on March 24, 1494 and died in Chemnitz, Saxony on November 21, 1555. The gifted Agricola (Fig. 8.1) studied classics, philosophy, and philology at the University of Leipzig 1514−18 where, after 3.5 years, he attained the degree of *Baccalaureus Artium*. Georg Bauer Latinized his name to *Georgius Agricola*, by which he was known throughout his professional life. At age 24, he was appointed *Rector extraordinarius* of the Greek School of Zwickau (1518−20), where he taught Latin and Greek.

After 2 years in Zwickau, Agricola gave up his position at the Greek School to become lecturer at the University of Leipsic, working with his friend, Professor of classics Petrus Mosellanus (1493−24). In Leipsic, he also studied *medicine, physics,* and *chemistry.* He received encouragement from Mosellanus, a famous humanist of the time. Agricola went to Italy in 1524, visiting the universities of Bologna, Venice, and Padua, where he completed his medical education in 1524−26 and studied philosophy. During his stay in Italy, he met *Erasmus van Rotterdam*, the great Dutch humanist and theologian, at that time editor for The Froben Press in Basel, and who became his lifelong friend.

Agricola returned to Zwickau in 1526 and in 1527, he was chosen as town physician in Joachimsthal, Bohemia—a booming place and a center of mining in the Erzgebirge (Ore Mountains), a mountain range separating Saxony and Bohemia. He devoted the time outside of his medical duties to visiting the mines and smelters (Ger = Hütten)

Toxicology in the Middle Ages and Renaissance. DOI: http://dx.doi.org/10.1016/B978-0-12-809554-6.00008-1

*Figure 8.1 Statue of Gregorius Agricola' in a small park in front of the railway station of Glauchau, Saxony, Agricola's birthplace.* Photo: S Langård.

and aimed, through the study of ores and their treatment, to better understand and revise, as necessary, the science of mineralogy. His background in philosophy was useful in enabling him, beginning in 1528, to develop a logical and systematic structure for the science. The manuscript of Agricola's textbook *"Bermannus, sive de re metallica dialogus"* (1530) [*Bermannus; or a dialogue on metallurgy*] was submitted to Erasmus, who had it published by The Froben Press in 1530. Erasmus also wrote a warm letter of approval which appeared at the beginning of this "catechism" on mineralogy. In the early 1530s, Agricola started work on *De re Metallica*.

*Prince Maurice* of Saxony appointed Agricola as historiographer in 1530. He accordingly moved to Chemnitz, another important mining center in Saxony, allowing him to widen the range of his observations on minerals and mining. In 1533, he published a book about Greek and Roman weights and measures, *De Mensuis et Ponderibus*. The citizens of Chemnitz showed their appreciation by appointing him town physician in 1533, and in 1546, Maurice appointed him Burgomaster of Chemnitz, a position that he held for four terms.

## 8.2 DE RE METALLICA

Agricola is frequently referred to as "the father of mineralogy," a moniker largely resulting from his best known book, *De Re Metallica* (Zurich and Deventer, 1556); Agricola completed its 12 volumes in 1550, but it was not printed until 1556 when woodcuts for the numerous drawings were finally completed. In this compendium, he cataloged his observations of metal-containing minerals, and methods of mining, refining, and smelting.

He also refined the age-old mining technique of fire-setting in which flammable substances were set on fire against the rock, followed by quickly lowering the temperature with water, thus causing it to crack. Agricola reported on the toxic gases and health hazards generated from smoke during below ground *firesetting* and suggested ways to ventilate the gases (book VII, p. 214–18). The use of powder to blast rock ultimately replaced fire-setting. Agricola also suggested remedies for the cold and wet conditions often present in mines.

In an appendix to the book, German equivalents were listed for the technical Latin terms used in the body of the work. Modern words that derive from the Agricola work include *fluorspar* and *bismuth*. The book remained an authoritative text on mining for 180 years and was reprinted in at least 10 editions.

The first translation into English of *De re Metallica* was privately published in 1912 in London by Herbert Hoover, mining engineer, and later President of the United States, and his wife, Lou Henry Hoover, a geologist and Latinist. Their translation was prized not only for its clarity, but also for its extensive footnotes. Subsequent translations into German and other languages owe much to Hoover's English version, as their footnotes detail their difficulties with Agricola's invention of several hundred Latin expressions to cover medieval German mining and milling terms that were unknown to classical Latin (Hoover and Hoover, 1950).[a]

## 8.3 DISEASES IN MINERS AND THEIR PREVENTION

Another important topic covered in *De re Metallica* were the possibly toxic effects of certain heavy metals associated with silver mining, e.g., cobalt and arsenic. Cobalt was a by-product of silver mining, while arsenic was present in *pompholyx*—a furnace deposit *"which eats wounds and ulcers of the bone."* Elsewhere in the book, Agricola notes toxic effects typical for arsenic.

The most remarkable revelation in the textbook was on respiratory diseases among miners, which he attributed to exposure to dust: it *"penetrates into the windpipe and lungs, and produces difficulty in breathing, and the disease which the Greeks call* ἄσθμα." (or asthma, and which was called Bergsucht in German), and further *"it eats away the lungs, and plants consumption in the body."* Agricola made no mention of having performed autopsies on deceased miners, but it seems likely that he did,—given his observation that the disease, whatever it actually was *"eats away the lungs."* *Bergsucht* might have been *pneumoconiosis* or *tuberculosis*, both traditional diseases of miners. The text does refer specifically to lung cancer—a localized disease that does not *"eat away the lungs"*—as it was likely not understood at the time (Hoover and Hoover, 1950).

Agricola reports (p. 214) on the possible hazardous effects of the mineral *cadmia* *"which eats away the feet of the workmen when they become wet, and similarly their hands, and injures their lungs and eyes."* In the German translation, *Kobolt* (or *cobaltum* by Agricola, p. 214), was probably arsenical-cobalt, another mineral common in the Saxon mines and which Agricola made note of. Therefore, Agricola advises the miners (p. 215) to *"make for themselves not only boots of rawhide, but gloves long enough to reach the elbows, and they should fasten loose veils over their faces; the dust will neither be drawn through these into the windpipes and the lungs, nor will it fly into their eyes."* He further revealed that the Romans, when making the mercury-containing *vermilion*, took precautions against breathing its fatal dust. (Vermilion is an opaque pigment, originally made by grinding *cinnabar ore* containing mercury sulfide.)

---

[a]Denotation of the pages in this chapter refer to the 1950-edition by Hoover and Hoover.

On p. 217, Agricola refers to "*some of our mines, however, though in very few, there are other pernicious pests. These are Demons of ferocious aspects, about which I have spoken in my book De animantibus subterraneis. Demons of this kind are put away by prayers and fasting.*" Agricola seems to accept the presence of demons or gnomes in the mines, which was a general belief at the time. This may seem strange, considering the author's general skepticism about the supernatural.

In volume IX of *De re Metallica*, Agricola also presents descriptions of the melting of metals and methods for separation of gold and silver from other metals through the process of amalgamation, i.e., creating an alloy of mercury and another metal (p. 297–298). However, he makes no mention of hazards from the mercury fumes. The use of mercury for amalgamation of gold is a method that was reported on by Pliny the Elder. However, this method clearly results in toxic effects from lead fumes.

Agricola attributed extensive fish kills in rivers to pollution by metals and minerals. He pointed out the hazardous effects of water pollution from toxic elements. Perhaps, an environmentalist at heart, he was concerned that razing forests for mine construction and ore smelting, would lead to the eradication of birds in Saxony. Agricola was preoccupied with preventing accidents, and he underlined the need to prevent falling rocks and tunnel cave-ins; "*therefore, miners should leave numerous arches under the mountains which need support, or provide underpinning.*"

## 8.4 *BERGSUCHT* AND ITS CAUSES

Agricola did not attempt to identify the possible causes of *Bergsucht*. Over the following centuries, though, light was shed on its causes.

On August 18, 1785 Johann Wolfgang von Göthe traveled from *Karlsbad* to *Johanngeorgenstadt* (currently in the Czech Republic) to collect rock samples. One of the minerals was named "Pechblende". Four years later a sample of *Pechblende* was sent to the pharmacist Martin H. Klaproth in Berlin for analysis. He identified a new element in the sample that he named *Uran*. Recovery and processing of uranium from *Pechblende* for glass manufacture and for luminescent material began in 1820. This will soon play a role in our story.

In the village of *Schneeberg*, mining for *copper* and *silver* started about 1410 (Hueper, 1942) and continued for five to six centuries. By the late 19th century, there was a complex of six mines employing on average some 650 miners who extracted ore largely for the production of nickel, cobalt and bismuth. Härting and Hesse (1879a–c) reported on an *"endemic occurrence"* of *lung cancer* among the Schneeberg miners, of whom 75% died prematurely. They established a cohort of miners from the Erzgebirge range in 1879 and reported a very high occurrence of lung cancer in this population. Among this population of 650, they recorded 63 deaths during 1869–71, 42 during 1872–74, and 40 during 1875–77.

The authors reported that pulmonary malignancy was the cause of 75% of the deaths. They also observed that 20 years exposure to mine conditions was required for lung cancer to develop, and that, rock cutters developed the disease more quickly than any other group when they perform under-ground work on a continuous basis. In the course of their study, Härting and Hesse made enquiries to physicians at similar mines at Modum (Norway), Joachimstal (Bohemia), Dobschau (Hungary), and Leogang (Tyrol), and confirmed that lung cancer had not occurred at these sites.

In one publication, Härting and Hesse (1879c) state: *The endemic miners' disease that occurs in the Schneeberg mines is primarily lung cancer, and about 75% of the cases of death are caused by it. The following facts were also established*: (1) all miners not dying of other causes, eventually died of lung cancer, (2) the onset of the disease can occur anywhere to 20 to 50 years after working in the mines.

Härting and Hesse (1879a–c) also found that the disease named *"Bergsucht"* in the Schneeberg mineworkers actually was *lung cancer*. Subsequently, in 1913, radioactivity in the mine air due to radon was identified as the likely cause of the miners' lung cancers (Ludwig and Lorenser, 1924). Hence, the lung disease observed by *Agricola* is likely to have been lung cancer after all, this being the first work-related lung cancer ever identified.

## 8.5 A THIRD WAVE OF MINING IN THE 20TH CENTURY

Agricola first reported on the metal bismuth in 1520. The Germans mined bismuth (Ger = Wismut) during the first half of the 20th century, using it for pigment manufacture as well as in drugs. Meanwhile, in June 1946, a Soviet company named *"SDAG Wismut"* was established to exploit the major uranium deposits in Saxony. It may have been convenient to use the name Wismut AG, at least temporarily, to hide the real purpose of the mines. In the wake of World War II, research on and exploitation of atomic power was prohibited in Germany. Thus, uranium mined by Wismut AG was transported to the Soviet Union, which had a great need of the chemical and where only minor reserves were known (Schiffner, 1994).

The 600-year history of mining in the Erzgebirge was about to stop when this uranium ore mining was initiated after World War II. Qualified workers and adequate mining equipment were scarce. However, there were few new work places in Germany after World War II. Therefore, the Soviet administration could order the employment centers to open mines and, in a short time, thousands of people were drafted to work in uranium mining. By the end of the 1940s, more than 100,000 workers were employed by Wismut AG, but the work force was gradually reduced to 42,000–43,000 by the mid-1970s. In all between 400,000 and 500,000 workers were employed at Wismut AG between 1946 and 1990, mining a total of 251,510 t of Uranium.

Initially, workplace protective measures were inadequate, resulting in elevated risk of disease among the miners. The drill hammers available in the late 1940s did not permit wet drilling—resulting in exposure to quartz dust for thousands of miners. Inadequate ventilation caused high concentrations of quartz dust and radon and its decay products in the mines, resulting in worker silicosis and lung cancer. The annual radon exposure above 450 *working level months*, later lowered to 200 WLM (WLM—a measure of occupational exposure to radon daughters) was required to qualify a case of lung cancer as occupational (Figure 6.3-1 in Kreuzer et al., 2011). As of 1990, 5200 cases of lung cancer have qualified as occupational cancer. Up to 1955, the yearly dose had been estimated at 150 WLM, while the yearly exposure after 1966 was estimated at 4 WLM/y. Kreuzer et al. (2011) reported on 3016 deaths from lung cancers through 2003— among about 18,000 deaths—in a subcohort of 59,000 workers, plus additional lung cancer cases in the subsequent years. Given that this subcohort is a good surrogate for the total cohort, the total extrapolated number of lung cancers through 2015 would be ± 15,000 deaths from lung cancer (Fig. 8.1).

Lung cancer had not been identified as a distinct disease during Agricola's time. Thus, he could not recognize work-related lung cancers. The *Bergsucht* that he reported on might have also represented other lung diseases such as pneumoconiosis or tuberculosis. However, the confirmation by Härting and Hesse (1879a–c) that the miners primarily developed lung cancer, strongly indicate that Agricola's *Bergsucht* was lung cancer, hence that, he also indirectly predicted the largest epidemic of work-related lung cancer ever encountered, that at the Wismut AG, Saxony—from the 1960s and onward.

Regrettably, Agricola did not end his days in peace. Though the population of Chemnitz had turned to the Lutheran creed, he remained a staunch Catholic to the end. It has been claimed that his life was ended by a stroke, brought on by a heated religious debate with a Protestant, and he died in Chemnitz in 1555. So violent were anti-Catholic attitudes that he was not granted burial in the town that he loved. Amidst a hostile crowd, his body was transported 50 km away to his final resting place in Zeitz.

## CONFLICT OF INTEREST

None.

## REFERENCES

Härting FH, Hesse W: Der Lungenkrebs, die Bergkrankheit in den Schneeberger Gruben, *Vierteljahresschrit für Medizin und Offentliche Gensundheitswesen* 30:296–309, 1879a.

Härting FH, Hesse W: Der Lungenkrebs, die Bergkrankheit in den Schneeberger Gruben. Vierteljahresschrift für Medizin und, *Offentliche Gesundheitswesen.* 31:102–132, 1879b.

Härting FH, Hesse W: Der Lungenkrebs, die Bergkrankheit in den Schneeberger Gruben, *Vierteljahresschrift für Medizin und Offentliche Gensundheitswesen* 31:313–337, 1879c.

Hoover HC, Hoover LH: *Gregorius Agricola De Re Metallica Translated from the First Latin Edition of 1556*, New York, 1950, Dover Publications, Inc.

Hueper WC: *Occupational tumors and allied diseases*, Springfield, III, 1942, Charles. C. Thomas.

Kreuzer M., Grosche B., Dufey F., et al.. Bundesamt für Strahlen-Schutz. Tecnical Report. The German Uranium Miners Cohort Study ("Wismut cohort"), 1946–2003. Oberschleissheim, February 2011.

Ludwig P, Lorenser S: Untersuchungen der Grubenluft in den Schneeberger gruben auf den Gehalt an Radiumemanation, *Z Physik* 22:178–185, 1924.

Schiffner W. "Agricola und die Wismut". Sachsenbuch Verlagsgesellschaft mbH. 1994; 1–127.

# CHAPTER 9

# Jan Baptist Van Helmont and the Medical–Alchemical Perspectives of Poison

**Georgiana D. Hedesan**
University of Oxford, Oxford, United Kingdom

In his masterpiece *Ortus medicinae* (1648), the Flemish polymath Jan Baptist Van Helmont (1579–1641) (Fig. 9.1) recounted his meeting with a mysterious Irish alchemist named Butler. Butler was in the possession of a "little stone," *lapillus*, a wondrous alchemical medicine that could cure any disease by touching it with the tip of one's tongue. Van Helmont was himself given this cure, which, he said, healed him of a slow poison given by an enemy, who confessed his guilt on his deathbed (Van Helmont, 1652, p. 469). As he pondered the *lapillus* and its contents later on, Van Helmont compared its action with that of viper venom, which also acted instantly and in very small quantity (Van Helmont, 1652, p. 474). Thus, he envisaged snake poison and Butler's *lapillus* as polar opposites. However, Van Helmont was not content with alchemy's action being just as powerful as that of poison, and was eager to affirm its superiority: alchemical medicine could overcome any type of disease or poison. This chapter analyzes how Van Helmont used the notion of poison and poison theories to legitimize the pursuit of medical alchemy. In doing so, he developed his ideas in light of those of Theophrastus von Hohenheim, called Paracelsus (1493–1541), the maverick Swiss physician who initiated a highly influential, if controversial, movement of medical alchemy.

## 9.1 VAN HELMONT'S PARACELSIAN LEGACY ON POISONS

Van Helmont was an heir of Paracelsus's thought and movement, even if, in his later years, he went to great lengths to distinguish himself from the Swiss physician. While in his early writings, Van Helmont had praised Paracelsus as the restorer of true medicine, by the time

Toxicology in the Middle Ages and Renaissance. DOI: http://dx.doi.org/10.1016/B978-0-12-809554-6.00009-3

*Figure 9.1 Portrait of Jan Baptist Van Helmont, from Aufgang der Arzney-Kunst (Sultzbach, 1683). Credit: Wellcome Library, London (CC BY 4.0).*

most of the *Ortus medicinae* was written (between 1637 and 1644), Van Helmont was keen to affirm himself as a original philosopher who alone had come into the possession of the true "Christian philosophy." Indeed, his wish was to create a new synthesis between Christian thought and natural philosophy, which would rely on the Bible and alchemy as the main pillars of its ideas (Hedesan, 2016).

Van Helmont was a keen reader of Paracelsus and had inherited his framework in regards to poison. As I have shown elsewhere, the Swiss physician had developed highly complex theories of poison (Hedesan, 2017). Perhaps the most important of them was his theory of universal poison, which argued that all things contained poison within them. This view was later refined by a "potent poison" theory, according to which some beings have more poisonous power within them than others. Both of these theories had a strong alchemical undertone, since they argued that only alchemy was able to remove the poison within things and hence provide powerful medicine for the sick.

By comparison to Paracelsus, Van Helmont did not develop extensive poison theories. However, as will be seen, he was influenced by the universal and potent poison theories of the Swiss physician. These took idiosyncratic forms as Van Helmont tried to reconcile them with his own specific worldview and speculation. Moreover, his perspective on poisons was deeply tied in with his sharp criticism of Galenic medicine and his supreme faith in the power of medical alchemy.

## 9.2 UNIVERSAL POISON IN A CHRISTIAN PERSPECTIVE

In his writings, Paracelsus had strongly advocated a Christian approach to philosophy and medicine, but his theory of universal poison seemed to stand at odds with it. How was it possible that a benevolent Christian God would permit the existence of poison in all things? Paracelsus attempted to answer this by formulating a complex theory of how things were good in their essence, but ambivalent in relation to other things (Hedesan, 2017). This was an ingenious theory, but did not completely solve the problem. Nor could he explain precisely why some things were more poisonous than others.

As a Christian medical alchemist and Paracelsian follower, Van Helmont faced a similar conundrum. The theory of universal poison was powerful and appealing because it legitimized medical alchemy, confirming the fact that it alone was able to separate the good from the evil in things. Van Helmont could not give up on it, but in order to advance it, he had to answer the same question that plagued Paracelsus.

In *Ortus medicinae*, Van Helmont did his best to clarify this problem. In a typical move, he expressed his theory in terms of a biographical quest. He had long observed that "if we investigate in depth, there is hardly anything in nature that does not have poison secretly mixed in itself" (Van Helmont, 1652).[1] Even roses and violets hid poison within, as all things contained impurity, residue, or crudity. We can recognize in this an enunciation of Paracelsus's universal poison theory.

---

[1] All translations from Latin are my own.

As a good Christian, Van Helmont found himself wondering what the situation was before the Fall of Man itself. He considered that poisons could not harm Adam before the Fall, because he was immortal by the Tree of Life; Van Helmont also pondered that perhaps, the snake notwithstanding, Paradise was free of them. Poisons could only be found in the ordinary world outside of Paradise.

This account raised several issues—Why were there so many poisons in the (nonparadisiacal) world in the first place? And how could he reconcile the universal poison view with the fact that the Bible clearly said that everything was made good in itself? Van Helmont confessed that he had grappled with these questions for a long time.

Luckily, he maintained, alchemy came to the rescue and shed a ray of light. In the laboratory, Van Helmont discovered that poisons could be changed by "our small labor" (parvo nostri studio) into effective medicine. Even more, importantly, the more horrid the poison was, the more powerful the remedy. This latter view concurred with Paracelsus's "potent poison" theory.

Thus, poisons were not ultimately evil, but good in their essence. Van Helmont went further: he postulated that God was not responsible for death, disease or evil; man had created these because of his Original Sin. Van Helmont envisaged the Fall as having dramatically damaged the constitution of Adam. It was not that poisons were harmful, but that the human body became imperfect and could be poisoned. Thus, the Fall did not alter the ontological goodness of things, but many poisons, due to their natural strength, were subsequently able to cause damage to the inferior bodies of human beings.

Van Helmont concluded that goodness lay just below the poisonous husk (siliqua), a fact which was confirmed, he maintained, by a chance happening. Working on wolfsbane (aconitum) one day, he touched the alchemical solution he had prepared with the tip of the tongue. He soon felt his head heavy and eventually fell into a strange state of stupor. He had the sensation that he understood everything very acutely, and considered that this clarity was located at the top of the stomach. This gave him the conviction not only in the goodness of apparently evil things, but also in the fact that there was great power in poisons, if only their venomous side were removed.

Furthermore, he thought that the Bible confirmed this, since Ecclesiasticus 38:4 stated that God had created medicine out of the earth.[2] Van Helmont also pondered that the goodness of God must always be stronger than illness and disease. This was already seen, he noted, in the swift and effective action of a famous antidote against snakebite called Orvietan (Van Helmont, 1652; on Orvietan, see Catellani and Console, 2005). Yet, it was much more prominent in the *lapillus* of Butler, which cured all diseases by touching it with the tip of the tongue, or by applying dermally. Van Helmont had no doubt that alchemy held the key to true medicine, and contrasted the alchemical processes with the Galenic approach to poisons.

## 9.3 VAN HELMONT'S CRITICISM OF GALENIC PURGATIVES

In Van Helmont's time, the majority of physicians followed the Galenic framework, which at this time incorporated medieval Islamic and Latin developments.[3] Galenic medicine was based on the doctrine of the four humors (blood, yellow bile, black bile, and phlegm), which regulated the body. Disease arose when these bodily fluids were imbalanced, and the "balancing" act was done by tempering or purging the offending humor. Common treatments included taking medicines extracted out of plants (usually referred to as "simples") and bloodletting.

In *Ortus medicinae*, Van Helmont launched a withering attack on the Galenic pharmacopoeia. First, he noted that many plants had a faculty that was stronger than that of the human, thus becoming poisonous to the body. Consequently, most Galenic physicians either rejected them altogether, or sought to correct them by inappropriate means (Van Helmont, 1652). As he pointed out, correction by boiling did not just remove the poison but also the remedy. For instance, scammony[4] boiled or treated with acids lost its strength (Van Helmont, 1652) (Fig. 9.2).

---

[2]The whole passage runs: "The Lord hath created medicines out of the earth; and he that is wise will not abhor them." Ecclesiasticus (Jesus Sirach) is considered a Deuterocanonical book in the Catholic faith and apocryphal in the Protestant one. In the period, Ecclesiasticus was popular with medical alchemists, as they believed it justified their doctrines.
[3]On medieval medicine and early Renaissance medicine, see Siraisi (1990).
[4]Scammony (*Convolvulus scammonia*) is a perennial plant native to the countries of the eastern part of the Mediterranean basin; it grows in bushy wasteland from Syria up to Crimea, its range extending westward to the Greek islands. The juice of scammony has a powerful purgative effect, see Chisholm (1911).

*Figure 9.2 Image of "True Scammony" from Medical Observations and Inquiries (London, 1757).*
Credit: Wellcome Library, London (CC BY 4.0).

Such boiled "medicines" were ineffective, but at least they were harmless; by comparison, other Galenic corrections were downright dangerous for human health. Van Helmont gave the example of the physicians' erroneous treatment of the Spanish—Italian general Carlo Spinelli, who was given a solution of white hellebore corrected with anise seed. This provoked a half-hour vomiting bout that ended in convulsion and ultimately death (Van Helmont, 1652).

Van Helmont considered that the problem with Galenic physicians was that their attitude toward hard purgatives, particularly laxatives, was inconsistent and hypocritical. While Galen had thought of laxatives to be poisonous, and most Galenic physicians would admit that they were dangerous, they continued prescribing them in "corrected" form. When things did go wrong, they blamed the dose, the correction, the solution, the apothecary or even their own wife for it (Van Helmont, 1652).

The attraction of Galenic physicians to purgatives was borne out of a fundamental error: they falsely believed that excrements induced by purgatives were bad humors. Van Helmont denied that the excrements resulting from laxatives were humors at all, but putrefied matter caused by poisoning. Again, Van Helmont's evidence lay in personal testimony, told, in Helmontian style, with rueful humor.[5] When he was young, he recounted, he had touched the glove and hand of a lady infected with scabies, and got the scabies too. At the time, Galenic physicians did not recognize the disease as being contagious, but instead attributed it to a distemper caused by the overheating of the liver. Young Van Helmont, 18 at the time and a student of medicine, called in two leading physicians of Brussels to obtain their recommendation, "half-glad" to get experience in medical treatment. The physicians diagnosed him with an excess of inflamed bile and of salty phlegm, which caused damaged blood production in the liver. First, they prescribed bloodletting to cool the liver. Secondly, a concoction was offered to eliminate the yellow bile and phlegm from the body. Since that also did not work, Van Helmont took laxative pills of *fumaria*, which made him evacuate many stools. Naively, he felt pleased by this as he thought his corrupted humor was thus being eliminated. He took another round of laxatives after 2 days, and then another after three. Having eliminated so many stools as to "easily fill two buckets," Van Helmont felt seriously weak: "I who previously were healthy, vivacious, full of strength, light in jumping and running, I was now emaciated, my knees were trembling, my cheeks collapsed, and my voice was hoarse" (Van Helmont, 1652). Nor did his scabies go away with the laxative. At that point, the youth started wondering where all those humors were coming from. Surely, he calculated, there was not enough room in the bowels, in the head, and in his chest to contain so much humor. Eventually, upon reflection, Van Helmont concluded that the so-called humors were not originally present in his body but were formed by the action of the laxative. This conviction acquired, the young man grew increasingly critical of traditional medicine. As for laxatives, he concluded, "it is indubitable that [they] contain a hidden poison, which has made thousands of widows and orphans" (Van Helmont, 1652).

Eventually, Van Helmont claimed that laxative action was simply a manifestation of their poison. The mechanism whereby poisons acted within the body was the following: once ingested and received in the

---

[5]The story is first told in the treatise "De febribus," p. 756 (published first in 1642), and later in "Scabies et ulcera Scholarum," pp. 255—256.

stomach, they fermented, dissolved anything found there, and then putrefied them. This would still be fine if the laxatives were to dissolve excrements, but they actually destroyed the vital juices of the body. Thus, they corrupted the purified blood out of the *vena cava*, contaminated it with their poison, and dissolved it by means of a fetid cadaverous ferment (Van Helmont, 1652). This led to disturbance of the body that persisted long after the laxative was taken, and often in spite of any astringent medicine. Hence, he concluded, laxatives "are poisons to us, not to the excrements" (Van Helmont, 1652).

Van Helmont was also critical of the common belief in his period that laxatives could "clean" the body and supposedly preserve it from disease. He gave the anecdotal example of a Privy Counselor of Brabant, who took aloe pills to maintain his health intact. These pills were corrected to the point they were useless, and had no effect. When the counselor complained of the lack of success to his physician, the latter gave him stronger pills of an undisclosed type. These had such a harsh purgative effect that the counselor died miserably, leaving behind 11 children (Van Helmont, 1652). Van Helmont blamed the faith in laxatives squarely on Galenic practices, and on the false belief in humors.

## 9.4 THE ALCHEMICAL SOLUTION

As already mentioned, Van Helmont believed, along with Paracelsus and other Paracelsians, that the essence of all things was pure and good. This belief was articulated in Van Helmont's theory of the *primum ens* ("first being"), which was drawn from pseudo-Lullian and Paracelsian precedents. In Van Helmont's view, if an entity could be returned to its *primum ens*, it could reacquire its original goodness, or medical quality. This was particularly important in the case of poisons, which had a stronger virtue than others.

Van Helmont expressed the alchemical process of obtaining the medical essence as an "inversion." A poison should be "inverted in its core" (*in sui radice introverti*): colocynth, for instance, could "invert" its laxative quality, which could then be used to cure chronic disease (Van Helmont, 1652).

This, Van Helmont added, was familiar to Paracelsus, who knew how to accomplish the inversion for a medicine called antimonial tincture of lily. Yet, the Flemish physician maintained, Paracelsus did not know that this could be done for all poisonous plants and animals by using the "greater circulated salt" solvent. Indeed, all things lost their poison and acquired medical power if they were reduced to their *primum ens* (Van Helmont, 1652).

The "greater circulated salt," the same or similar to the universal solvent *alkahest*, emerged as the primary substance capable of bringing about the medical inversion of poisons.[6] Van Helmont eloquently praised this alchemical key as that "which returns all things into the *primum ens*, preserving their native endowments, erasing the original blemishes of bodies; once their inhuman ferocity is removed, they become capable of giving birth to great and inexplicable powers" (Van Helmont, 1652). Ultimately, the *alkahest* could lead to the creation of the supreme universal medicne, Butler's *lapillus*, which demonstrated the special providence of God toward mankind.

Thus, alchemy provided the key to remove poison from all things. However, not all essences were equally useful for the human body. Van Helmont supported the use of poisonous plants in medicine, but was ambivalent about animal poison. He believed, along with most medical alchemists of his time, that metals and minerals ordinarily deemed poisonous, like antimony and mercury, could be exploited for great health benefits. Yet, he absolutely condemned the internal employment of arsenic in any form. Finally, he extolled the virtues of poisonous "sulfurs" of metals and minerals, which could be transformed into medicines that were particularly comforting to the body. Van Helmont listed many potentially deadly diseases healed by means of sulfurs (Van Helmont, 1652).

## 9.5 CONCLUSIONS

Van Helmont's interest in poisons was deeply linked with his belief in the power of medical alchemy. For the Flemish physician, alchemy was the apex of all knowledge and was most effective in providing cures to diseases previously—and erroneously—thought to be incurable.

---

[6]On the *alkahest*, see also Porto (2002), Hedesan (2016, pp. 177–182).

Undoubtedly, Van Helmont's understanding of alchemy derived from Paracelsus and Paracelsianism. This was particularly evident for poisons. Similar to Paracelsus, Van Helmont thought that poisons acted as a link between natural philosophy and medical–alchemical practice. They were tied with a grand vision of good and evil in the world, and with an attempt at explaining the presence of evil in Christian terms. They were also meant to justify peculiar and controversial alchemical practices.

We have seen that Van Helmont fundamentally subscribed to Paracelsus's universal poison theory, and the associated potent poison theory. Like Paracelsus, he believed that things had a "poisonous" and a "medicinal" side. He also believed the alchemy could separate these two aspects so that only the medicinal side remained. Finally, he also agreed that some substances have more power than others, and these hid great medicine within.

If the two men disagreed on anything, it was on the details of the philosophy behind the theory. Paracelsus sought universal theories that integrated man with nature. Van Helmont preferred to focus his attention on man exclusively, whom he found responsible for everything from death to disease and poison. Van Helmont's philosophy was essentially a human philosophy, targeted and justified by medicine.

## REFERENCES

Catellani P, Console R: Orvietan, a popular and controversial Panacea, *Pharm Hist* 35:11–19, 2005.

Chisholm Hugh, editor: *Scammony. Encyclopædia Britannica*, ed. 11, Cambridge, 1911, Cambridge University Press.

Hedesan GD: *An alchemical quest for universal knowledge: the "Christian Philosophy" of Jan Baptist Van Helmont (1579–1644)*, London, 2016, Routledge.

Hedesan GD: *Alchemy, potency, imagination: Paracelsus's theories of poison. In Cunningham A and Grell, OP: Poisons in European History*, London, 2017, Routledge, (accepted, in press).

Porto P: "*Summus atque felicissimus salium*": the medical relevance of the liquor alkahest, *Bull Hist Med* 76:1–29, 2002.

Siraisi NG: *Medieval and early Renaissance medicine: an introduction to knowledge and practice*, Chicago, 1990, University of Chicago Press.

Van Helmont JB: *Ortus medicinae, id est, initia physicae inaudita*, ed. 2, Amsterdam, 1652, Elsevier.

# CHAPTER 10

# Origin of Myths Related to Curative, Antidotal and Other Medicinal Properties of Animal "Horns" in the Middle Ages

**Chris Lavers**
University of Nottingham, Nottingham, United Kingdom

## 10.1 A BRIEF HISTORY OF ALICORN

In late Medieval and Renaissance times, alicorn was a sought-after and very expensive commodity in Europe (Shepard, 1930; Gotfredsen, 1999). In 1609, Thomas Decker compared the value of a single horn to half a city. In 1553, the King of France's alicorn was valued at £20,000. Half a century later, the British royal family's Horn of Windsor was valued at £100,000 (though perhaps this is an early example of a decimal-point error: A later reference states £10,000). Exaggeration aside, at its zenith alicorn, was worth 20 times its weight in gold, and even diced or powdered alicorn half that much (Shepard, 1930).

Consensus on the archetype of alicorn was reached in the 13th century: Spiral horns began to appear on unicorns in Christian art around the year 1200 and became the iconographic standard within a century or so (Lavers, 2009). This "horn" (it is a tooth) belongs to a large marine mammal, the narwhal. Narwhals frequent the seas around eastern Canada and Greenland, but occasionally wash up on beaches as far south as Britain and Germany. Narwhal tusks also traveled south as trade items. The seas around Greenland, via the Norse, supplied most of Europe's alicorns, though tusks from northern Eurasia probably ended up on the market too.

Some 12th-century Europeans came to believe that unicorn body parts had medicinal virtues. By the end of the 14th century, this belief had become attached to alicorn, as had the idea that alicorn could detect poisons (Shepard, 1930). Between the 15th and 17th centuries, alicorn was regularly used by the wealthy as a defense against

Toxicology in the Middle Ages and Renaissance. DOI: http://dx.doi.org/10.1016/B978-0-12-809554-6.00010-X

poisoners: It was believed to perspire in the presence of adulterated food and drink (it doesn't). From the late 16th century, poorer people entered the market for alicorn powder, sold for the treatment of numerous ailments. The constituents of powders being difficult to confirm, powdering was the key to tapping the dispersed wealth of ordinary people, and business boomed.

The alicorn bubble was punctured in the 1630s but took over a century to deflate. Copenhagen merchants asked Ole Wurm, Regius Professor of All Denmark, to identify alicorn. Wurm stated flatly that it was the tusk of an animal that lived in the waters of the North Atlantic. Confronted with a choice between truth and profit, the merchants of Copenhagen carried on trading. Nevertheless, the craze for alicorn gradually faded through the 17th century and eventually fizzled out (Shepard, 1930; Lavers, 2009).

## 10.2 ORIGINS

From where, whom, and what did Europeans get the idea that alicorn had protective and curative powers? This is not an easy question to answer owing to the fragmentary nature of the literary record. The earliest Medieval source is *Physica* by Hildegard of Bingen (1098–1179). Having first reworked some Christian allegory, Hildegard offers her understanding of the unicorn's medicinal virtues (Miller, 1960):

*Pulverise the liver of a unicorn, give this powder in fat prepared with yolk of egg and make a salve, and there will be no leprosy... The leprosy comes of course oftentimes from the black bile and from the black stagnant blood.*

*If you make a girdle from the hide of the unicorn and gird yourself with it, no plague however severe and no fever will harm you. Also if you make shoes from the hide and wear them, you will always have sound feet, sound legs and sound joints, and also will no pestilence harm you while you are wearing them.*

Her instructions are added to by a later hand:

*If a man is afraid of being poisoned, he should place the hoof of a unicorn under the plate on which the food is or under the cup which contains his drink, and if there is poison in it, they become boiling hot if they are warm, but if they are cold they will begin to steam and thus will he know that poison is mixed therewith.*

Alicorn is not mentioned. Hildegard says elsewhere that under the unicorn's horn lies reflective metal, so she knows *of* alicorn, but seemingly not of any properties connected with it deserving of mention. News of alicorn's power then, seemingly, arrived and spread across Europe over the next two centuries (Shepard, 1930).

Where did these supra-Hildegardian notions from? Readers familiar with the unicorn's pagan roots might suspect the works of the Ancients, especially of Ctesias (c. 400 BC) and Aelian (c. 200 AD), in which the one-horned ass, one of the unicorn's literary ancestors, is credited with medicinal and antidotal powers (Lavers, 2001). However, both Aristotle and Pliny used Ctesias as a source, yet neither attributes strange qualities to the ass's horn. Pliny in particular would surely not have omitted such information had he known it. Just as peculiar, medical authors of Greek and Roman times, most notably Hippocrates and Galen, make no mention of the therapeutic properties of one-horned animals. The Bible and the Latin *Physiologus* (a medieval book of allegorical stories mostly about animals: see Lavers, 2009) are silent on the matter too. A tale linking the unicorn with poison does exist in the eastern lineage of *Physiologus* texts, but it seems to have been unknown in the West at the relevant time. As argued by Godbey (1939), the antitoxic and medicinal properties of the one-horned ass may have been added by a later copyist, who perhaps thought the Ancients had tried to describe a rhinoceros, whose horn was commonly associated with such properties in later writings (see below). True or not, there is no convincing connection, and much that argues against one, between the Ancients' writings on unicorned animals and the development in Medieval Europe of alicorn's reputation as a medicine and detector of toxins.

With such doubts in mind, the unicorn scholar Odell Shepard turned his suspicions eastwards. In *The Lore of the Unicorn* (1930), he makes his case by mentioning a few eastern ideas about unicorns before passing sentence on Medieval Arab writers and moving on. Shepard was knowledgeable and assiduous and not at all renowned for glossing over a subject, so perhaps he foresaw where an investigation of such matters might lead.

## 10.3 UNICORNS OF THE ARAB WORLD

In Hildegard's time, translations of Arabic texts were filtering into Europe. Is it merely coincidence that medicinal alicorn appeared when this transfer of knowledge was taking place?

Several one-horned animals roamed Medieval Arabic texts. Al-Biruni, a polymath of the late tenth and early 11th centuries (d. 1048), describes, for example, the karkadann in his book on India (See Ettinghausen, 1950 for quotations and a detailed treatment):

> It is of the build of a buffalo, has a black, scaly skin, a dewlap hanging down under the chin. It has three yellow hooves on each foot, the biggest one forward, the others on both sides. The tail is not long. The eyes lie low, farther down the cheek than is the case with all other animals. On the top of the nose there is a single horn which is bent upwards...

A further fragment of al-Biruni's work adds that the horn is conical, bent backwards and longer than a span, that the animal's ears protrude like a donkey's, and that its upper lip forms into a finger-shape. There is no doubt that this is the Indian rhinoceros, about which al-Biruni learned when he was in India. Al-Biruni was one of very few Medieval writers who knew the rhinoceros, however, which is why the karkadann subsequently diversified into several literary lineages of mythical animals. The gulf between the rhinoceros and the karkadann was widened when artists began drawing karkadanns with long, straight horns, rather than short, bent ones, and further still when they moved the horn from the rhino's nose to the karkadann's brow (for no reason anyone can divine). In the 16th century, rhinoceroses became better known and representations of karkadanns become increasingly recognizable as rhinoceroses, but in the critical time period of interest, the karkadann was routinely represented as a quadruped with a long horn sticking out of its forehead. This horn had become linked with poisons by the end of the 13th century. An early reference is made by al-Qazwini (d. 1283), who says that the substance is used both as an antidote and in the manufacture of knife handles (Ettinghausen, 1950), and later writers add that the horn perspires in the presence of poison. Here we seem to have the unicorn's geographically distant but genealogically close relative.

Did the karkadann inherit its powers from the unicorn or vice versa? A clue lies in the writings of al-Qazwini. Karkadann horn was used in the manufacture of knife handles, in which respect it was similar to another substance, khutu, with a longer literary history and a reputation for reacting in the presence of poisons. One suspects that karkadann horn inherited its reputation from khutu when both substances sat side by side in the workboxes of Arab cutlers.

Khutu was used as a poison detector and antidote in the Muslim world from the first half of the 11th century at the latest, because al-Biruni describes the material and its uses. Hildegard composed *Physica* in the middle of the 12th century, at least a century after al-Biruni's report on khutu, and by this time, European alicorn does not yet appear to have been linked with poisons. Absence of evidence is paradigmatically tricky, but the early date of tales about antidotal horns in the Arab world suggests that the mythological tradition traveled with the dominant flow of ideas then filtering from the Arab world into Christian Europe.

The reader may now have an inkling why Odell Shepard chose not to hunt the source of alicorn's power outside Christian Europe, but the foregoing is merely straightforwardly obscure. The natural historian cannot be content with 'khutu' as an answer to the likely source of alicorn's powers, but must ask the obvious next question: What is khutu?

## 10.4 KHUTU

The identity of khutu is a mystery that is a thousand years old and counting (Lavers and Knapp, 2008). A horn-like substance that reacts to poison brings to mind rhinoceros horn (Ettinghausen, 1950; Chapman, 1999), but in Arab literature, rhinoceros horn did not attract antidotal properties until the late 13th century (Ettinghausen, 1950), and probably gained them from khutu, the full transfer taking until the mid-15th century when rhinoceros horn was credited with khutu's characteristic reaction of sweating (Ettinghausen, 1950).

After extensive investigations, Laufer (1913, 1916) and Ettinghausen (1950) came to definite conclusions about what kind of material khutu was. In what follows I assess their claims, before dealing with some aspects of the khutu literature that they ignored or overlooked. Six descriptions of khutu are presented below in order to give an idea of the literature on khutu that Laufer and Ettinghausen were confronted with (a seventh, unknown to either investigator as he worked on the problem, will be forthcoming later). Synonymy of the words khutu, chatuq, khatu, khatuq, habaq, khataq, and other variants was demonstrated by Dankoff (1973).

Al-Biruni's description of khutu in an 1121 text by al-Khazini (Laufer, 1913) reads:

> It is asserted that it is the frontal bone of a bull living in the country of the Kirgiz who, it is known, belong to the northern Turks...The Bulgar bring from the northern sea teeth of a fish over a cubit long. White knife hafts are sawed out of them for the cutlers. The middle portion is distributed among the single hafts, so that every piece of the tooth has a share in them; it can be seen that they are made from the tooth itself, and not from ivory, or from the chips of its edges. The various designs displayed by it give the appearance of wriggling.

Elsewhere in the same work, al-Biruni adds information on khutu's shape and color (Laufer, 1913):

> It originates from an animal; it is much in demand, and preserved in the treasuries among the Chinese who assert that it is a desirable article because the approach of poison causes it to exude. It is said to be the bone from the forehead of a bull. Its best quality is the one passing from yellow into green; next comes one like camphor, then the white one, then one colored like the sun, then one passing into dark-gray. If it is curved, its value is a hundred dinar at a weight of 100 drams; then it sinks as low as one dinar, regardless of weight.

Ibn al-Husayn Kashghari's 11th-century definition of chatuq (Ettinghausen, 1950) runs:

> Horn of a sea fish imported from China. It is said that it is the root of a tree. It is used for knife handles. The presence of poison in food is put to the test by it because when broth or other dishes in the bowl are stirred with it the food cooks without fire, or if the horn is placed on a bowl it sweats without steam.

A text on precious stones by Ibn al-Akfani (d. 1348) includes a passage derived from al-Biruni's work (Laufer, 1913):

> Chartut is also called chutww...al-Biruni says: it originates from an animal. It is said to be obtained from the forehead of a bull in the region of the Turks in the country of the Kirgiz, and it is said also that it originates from the forehead of a large bird which falls on some of these islands; it is a favorite of the Turks and with the Chinese...The Ichwan al Razijans state that the best is curved, and that it changes from yellow into red, then comes the apricot-colored one, then that passing into a dust color and down to black.

A Chinese work, Cho Keng Lu, compiled in 1366 by T'ao Tsung-I, reads (Laufer, 1913):

*Ku-tu-si is the horn of a large snake, and as it is poisonous by nature, it can counteract all poisons, for poison is treated with poison. For this reason it is called ku-tu-si.*

A passage in al-Ghaffarı's mineralogy of c. 1511 reads (Laufer, 1913):

*The hutu is an animal like an ox which occurs among the berber and is found also in Turkistan. A gem is obtained from it; some say it is its tooth, others, it is its horn. The color is yellow, and the yellow inclines toward red, and designs are displayed in it as in damaskeening. When the hutu is young, its tooth is good, fresh and firm; when it has grown older, its tooth is also dark-colored and soft.*

There are other examples of this kind of description (Laufer, 1913, 1916; Ettinghausen, 1950; Dankoff, 1973; Said and Harmeneh, 1973; Said, 1989), but the problem of placing khutu in the natural world should already be apparent. Khutu is allegedly a horn, a tooth, a bone, and the root of a tree. It is derived from a bull or ox, a bird, or a snake, while something resembling khutu, though clearly not, it is owned by a fish.

## 10.5 THE WALRUS AND THE NARWHAL

Laufer (1913, 1916) worked mainly on the Chinese branch of the khutu problem and Ettinghausen (1950) on the Arabic branch, yet both concluded from their respective literatures that khutu was most likely ivory from the walrus, and perhaps also the narwhal. Walrus tusks are unique in the animal kingdom. The osteodentine core of a walrus tusk has a granular texture, like rice pudding. Al-Biruni's reference to the middle portion of the fish tooth from the northern seas implies that this core differs from the material surrounding it, and he also states that this core cannot be confused with ivory, presumably from an elephant, implying that the material on the outside of a walrus tooth could be so confused. Walrus ivory alone fits this description.

Other lines of evidence that khutu was walrus ivory are best seen in a text that was unknown to Laufer in the 1910s and was discovered by

Ettinghausen only as he was finishing his 1950 treatise and may be the original on which later textual fragments were based. In al-Biruni's Kitab al-jamahir fi mac rifat al-jawahir is the following passage (Said, 1989):

> ... When I enquired about the khutu from the members of the diplomatic mission which had come from the Qata'i Khan, they said: "The only merit about it is that it lets out perspiration when any poison comes into contact with it...It is the bone of [the] forehead of [a] bull." This is what has been said in books, although the only additional information we could get is that this bull is found in Khirkhiz. Its forehead is thicker than two fingers which would show that it cannot be the forehead of the Turkish bull, as it is smaller bodied. But it could well be the horn. As for the belief that it is the forehead of a bull, it would be the forehead of the mountain goats of Khirkhiz. Only they can have such foreheads ...
>
> It has patterns described over it and bears resemblance to the pith of the teeth of the fish which the Bulgarians bring to Khwarazm [Khiva] from the North Sea which is adjacent to the ocean. It is bigger than the hand in size and the pith is longer in the middle ...
>
> A Khwarazmian happened to find a tooth which was very white on the sides. He had hasps of daggers and knives made from it. The natural patterns described upon it were very thin, white and pale. It resembled the down of a cucumber if peeled in such a manner that that the seed grains are also cut off ...
>
> A tradition which runs about it—and it is extremely difficult to check the veracity of the factual truth behind this tradition—has it that it is the forehead of a big bird .... [Natives in the wilderness of China] believe it to be a very large fowl residing in uninhabited regions beyond the sea of Zanj and China, eating large ferocious elephants ...
>
> Amir Abu Jacfar ibn Banu had a large box-like case made of long and broad khutu planks....

Perhaps, the clearest evidence that khutu is walrus ivory concerns cucumbers: Pare this fruit lengthways with a vegetable peeler and toward the middle a pattern of translucent seeds set in an opaque matrix appears which closely resembles osteodentine (though the osteodentine I own has much smaller "seeds"). This pith is surrounded by cucumber of uniform appearance, just like the dentine surrounding the core of a walrus tusk. The case for the walrus thus seems secure, but other aspects of al-Biruni's text suggest that that is not the end of the matter.

## 10.6 THE WOOLLY RHINOCEROS AND MAMMOTH

Al-Biruni's avian allusion—"it is the forehead of a big bird"—may be rooted in stories about giant birds originating in Siberia, which may have been based on fossil skulls of woolly rhinoceroses (Laufer, 1916), though the motif may also be connected with the Arabic legend of the rukh (Ettinghausen, 1950), or the use by ivory carvers of the horny excrescence above the bill of the helmeted horn-bill (Laufer, 1916).

Mammoth ivory was also coveted by medieval Chinese craftsmen (Laufer, 1913). The theory that khutu was mammoth ivory was first proposed by Wiedemann (1991), and though Laufer (1913) tried to refute the idea he could not convince himself that the mammoth and khutu were unconnected. References to the Kirghiz in Arabic texts firmly underscored this suspicion. The Hudud al-'Alam of 982 states that Kirghiz lands supplied large quantities of khutu (Ettinghausen, 1950). The Kirghiz lived across a large area of Siberia between Arab and Chinese civilizations to the south and reindeer herders and marine mammal hunters to the north. Laufer (1916) conceded that trade relations between the Siberian Kirghiz and Arab lands hinted that "bulls furnishing ivory" may allude to mammoths and their tusks disinterred from the semi-frozen tundra of Kirghiz territory. Decades later, Ettinghausen (1950) agreed. As explained below, this "bull" may be a different kind of animal, but mammoth ivory may nevertheless have a connection with khutu. The reputed colors of khutu have always been a puzzle, routinely passed over by previous authors perhaps because fresh bones and teeth are white or yellowish. However, Siberian mammoth tusks may be variously colored, sometimes brightly so, depending on the chemical properties of surrounding rocks and soils. And if mammoth ivory was (a kind of) khutu, other puzzling attributions in the literature would make more sense. For example, cutting up the tusks of walruses or narwhals would hardly yield broad planks suitable for the construction of a large case (see above). Only ivory from an elephant or a mammoth seems to fit that bill. Since the Arabs knew about elephants, Banu's box of khutu planks may well have been made of mammoth ivory.

## 10.7 BONE FROM THE FOREHEAD OF A BULL

In a supplementary note, Ettinghausen (1950) comments, with obvious exasperation, on the text by al-Biruni quoted above that he discovered only as his work was going to press:

> Al-Biruni seems to imply that that he has actually seen khutu pieces. It is therefore significant that he distinguishes between the khutu and the fish tooth, i.e., the walrus tusk .... Unfortunately I cannot offer any satisfactory solution of the problem. It seems unlikely that al-Biruni is speaking of the tooth of the sperm whale .... The narwhal has no core with a design ... it is not likely that Egyptians would have paid a high price for hippopotamus teeth, which must have been fairly common in their country .... In case there is no other tooth like that of the walrus—nor a horn resembling it—the only remaining possibility would be that al-Biruni makes a distinction between two types of walrus teeth, perhaps teeth of different sizes or in different states of preservation; or we would have to assume that the cause of the whole confusion is of a semantic nature ....
>
> Why our author called the tusk a forehead bone remains another puzzle, especially since he himself preferred to call the khutu a horn.

This is where the matter was left in 1950. Pressing on, a material would seem to be required which fulfills some or all of the following criteria. It connotes the repeated description "bone from the forehead of a bull," but may be derived from a goat-like animal; it is hornlike, though sufficiently odd in appearance that al-Biruni was not sure what kind of horn it was; it is bigger than a man's hand (but presumably not bigger than, say, his forearm); it may have an internal structure that wriggles or damascenes, and in any case is suitable for the manufacture of knife handles; it is probably linked with the walrus, narwhal, and mammoth owing to its appearance, use, provenance, or association with Medieval trade routes.

One animal fits this template rather well. There are few species of large land animal in the far North, and only one that looks like a bull. We now know that musk oxen are more closely related to sheep and goats than cattle (Darwent and Darwent, 2004), but they have highly unusual horns which neatly fit the description "bone from the forehead of a bull." The horns of a male musk ox grow to form a continuous pad across its forehead, a boss, which is typically two or three inches thick. Underneath, the skull-cap underneath may be 4 in. thick (musk oxen butt heads hard). The musk ox would also make sense of al-Biruni's insistence that the piece of khutu he examined must have come from a goat (a head-butter) and not a bull (an impaler), because only goat-like animals have such foreheads. Al-Biruni was unsure if the khutu he examined was horn, but that is no surprise if he had been shown the grown-together mass of horny material that is a musk-ox boss; his hesitation would be a surprise, however, if he had been shown any regular kind of horn. And intriguingly, musk-ox horn is still one of a handful of substances that are especially prized by modern-day knife makers who specialize in using Arctic materials (Lavers and Knapp, 2008). Its value lies in its internal figuring, comprising intricate patterns of contorted—wriggling or damascening?—growth lines which form as the horns coalesce to form a boss.

Musk-ox horn thus seems to be a good candidate for the "bone from the forehead of a bull" that wriggles or damascenes (Lavers and Knapp, 2008). Indeed, if such a substance existed at all, no other high-latitude animal has the wherewithal to provide it. That said, one of T. H. Huxley's ugly little facts intrudes upon this otherwise elegant theory: Most authorities think the musk ox died out in Eurasia before al-Biruni's time. Even if that turns out to be true, however, it does not mean that musk-ox horn, still less stories about musk oxen themselves, could not have reached Medieval Arab cutlers.

It is becoming increasingly clear that Bering Strait has never presented much of a barrier to the movement of people (Fitzhugh, 1997), and a musk oxen roamed the Arctic slope of North America until the mid-19th century (they were wiped out and have subsequently been reintroduced; Allen, 1912; Lent, 1988). If musk-ox horns were mixed with walrus and narwhal tusks in bundles of Arctic goods, southern recipients may well have had cause to be confused about the distinction between a fish tooth and a bull forehead.

Another possible route around the ugly little fact leads to an established controversy about dating the extinctions of northern Eurasian animals. Until recently, it was thought that the musk ox died out in Eurasia at the end of the last ice age, but Russian mammologists have long suspected that musk oxen might have survived as isolated populations in Siberia long enough to encounter 17th-century Russian explorers (Spassov, 1991). Radiocarbon analyses of Eurasian musk oxen have produced ages of 3000–2700 years (MacPhee et al., 2002; see also Orlova et al., 2004; Boeskorov, 2006; Markova et al., 2015), much later than previously thought. Where the last Eurasian musk oxen lived is not known, so such radiocarbon dates may be considerably earlier than the true extinction date (this is the Signor-Lipps effect: We can never hope to find the last member of a species in the fossil record, so true extinctions always postdate the age of the youngest physical evidence). Intriguingly, in a burial complex in the Noin Ula Mountains of Mongolia, probably dating from the 1st century BC, two plaques were found impressed with an animal figure that looks very much like a musk ox (Spassov, 1991). Soergel (1942) suggested that the animal is a chimera, having the body of a takin and the head of a musk ox, the latter known to the artist from Siberian fossil skulls. But if Mongolians of the 1st century BC were familiar with such fossils, there seems little reason to deny that Chinese and Arab craftsmen of later centuries might also have been.

A different kind of explanation, arguably the likeliest, involves not the movement of materials but rather of tales about the animals that bore them. Stories about places and animals, as well as physical goods, clearly moved around the Medieval world. Tales of giant birds (woolly mammoth skulls?), or of bulls with outsized foreheads (Eurasian or American musk oxen?), might have traveled with variously colored fish teeth (fresh and fossil marine ivory and mammoth tusks) to be blended into a conundrum that has been confounding literary and natural historians for at least a thousand years.

## 10.8 CONCLUSION

Al-Biruni thought that khutu was the horn of a goat from the land of the Kirghiz (Siberia). Laufer (1913, 1916) thought it was marine ivory, though he suspected that mammoth ivory might have influenced the literary tradition. Ettinghausen (1950) concurred with most of Laufer's conclusions, though not always for the same reasons, but finished his work still puzzled.

Mammoth ivory may be wrapped up in the khutu enigma, but perhaps not in the way previously supposed. From mammoth ivory could have been derived ideas about khutu's coloration and its connection with cabinet-making. But al-Biruni's wriggling bone from the forehead of a bull does not fit comfortably with the walrus, narwhal or mammoth. Lavers and Knapp (2008) hoped to make it challenging to think of a better candidate for the source of this horn than the musk ox.

But whether musk-ox horn was a kind of khutu, or a story that kayaked across Bering Strait, or merely an oddly musky literary confusion, the geographical source of Medieval speculations about khutu now seems clear. The literature converges on a particular environment, place and people: Arctic Eurasia and its itinerant mammal hunters. Trade goods derived from various living and extinct Arctic animals probably became linked in people's minds and texts because they came from a mysterious place far to the north and made their way along the same trade routes. And eventually, it seems, one odd notion about these Arctic "horns" would circle back northwards and attach itself to an animal of very different provenance, Medieval Europe's fabled unicorn.

## DISCLAIMER

I offer here an overview of completed work rather than a guide to the literature; for the latter, please refer to a book that deals broadly with the natural history of the unicorn legend (Lavers, 2009), and two articles on ancient Greek and Roman (Lavers, 2001) and medieval (Lavers and Knapp, 2008) aspects of the natural history of unicorn lore. In the present piece, I use unaccented versions of Arabic names and words.

## REFERENCES

Allen JA: The probable recent extinction of the musk ox in Alaska, *Science* 36:720–722, 1912.

Boeskorov G: Arctic Siberia: refuge of the mammoth fauna in the Holocene, *Quaternary Int* 142:119–123, 2006.

Chapman J: *The art of rhinoceros carving in China*, London, 1999, Christie's Books.

Dankoff R: A note on khutu and chatuq, *J Am Oriental Soc* 93:542–543, 1973.

Darwent CM, Darwent J: Where the musk ox roamed: biogeography of tundra musk ox (*Ovibos moschatus*) in the eastern Arctic. In Lyman RL, Cannon KP, editors: *Zooarchaeology and conservation biology*, Salt Lake City, 2004, University of Utah Press, pp. 61–87.

Ettinghausen R: The unicorn. Freer gallery of art occasional papers 1 (3). Baltimore, 1950, Freer Gallery of Art.

Fitzhugh WF: Global culture change: new views of circumpolar lands and peoples, *Anthronotes* 19:1–8, 1997.

Godbey A: The unicorn in the old testament, *Am J Semitic Lang Lit* 56:256–296, 1939.

Gotfredsen L: *The unicorn*, London, 1999, Harvill Press.

Lent PC: Ovibos moschatus, *Mamm Species* 302:1–9, 1988.

Laufer B: Arabic and Chinese trade in walrus and narwhal ivory, *T'oung pao* 14:315–364, 1913.

Laufer B: Supplementary notes on walrus and narwhal ivory, *T'oung pao* 17:348–389, 1916.

Lavers C: The ancients' one-horned ass: accuracy and consistency, *Greek Roman Byzantine Stud* 40:327–352, 2001.

Lavers C: *The natural history of unicorns*, London, 2009, Granta.

Lavers C, Knapp M: On the origin of khutu, *Arch Nat Hist* 35:306–318, 2008.

Macphee RDE, Tikhonov AN, Mol D, et al: Radiocarbon chronologies and extinction dynamics of the late Quaternary mammalian megafauna of the Taimyr Peninsula, Russian Federation, *J Archaeol Sci* 29:1017–1042, 2002.

Markova AK, Yu Puzachenko A, van Kolfschoten T, et al: Changes in the Eurasian distribution of the musk ox (*Ovibos moschatus*) and the extinct bison (*Bison priscus*) during the last 50 ka BP, *Quaternary Int* 378:99–110, 2015.

Miller G: The unicorn in medical history, *Trans Stud Coll Phys Phila* 28:80–93, 1960.

Orlova LA, Kuzmin YV, Dementiev VN: A review of the evidence for extinction chronologies for five species of Upper Pleistocene megafauna in Siberia, *Radiocarbon* 46:301–314, 2004.

Said HM: *The book most comprehensive in knowledge on precious stones*, Islamabad, 1989, Pakistan Hijra Council.

Said HM, Hamarneh SK: *Al-Biruni's book on pharmacy and materia medica*, Karachi, 1973, Hamdard National Foundation.

Shepard O: *The lore of the unicorn*, London, 1930, George Allen & Unwin.

Soergel W: 1942 Lebten vor 2000 Jahrem Moschusochsen in der Nordmongolei. Naturund Volk 72: 41–55, 1942.

Spassov N: The musk ox in Eurasia: extinct at the Pleistocene-Holocene boundary or survivor to historical times, *Cryptozoology* 10:1–14, 1991.

Wiedemann E: Uber den Wert von Edelsteinen bei den Muslimen, *Der Islam* 11:345–358, 1991.

# CHAPTER 11

# Animal Stones and the Dark Age of Bezoars

**Maria do Sameiro Barroso**
Portuguese Medical Association, Lisbon, Portugal

## 11.1 INTRODUCTION

Throughout the ages, geological stones have been regarded, by widely dispersed civilizations, as a powerful symbol of infinite life and vigor. In the anonymous and apocryphal Hellenistic Lapidary *Orphei Lithica*, earth is called the mother of serpents and the mother of antidotes. It produces all kinds of stones of infinite strength and diversity. Plants may be powerful, but stones are even more so because the mother has provided them with a soul that neither dies nor gets old (Halleux and Schamp, 1985). Although bezoars are not minerals, they have ranked amongst the most prized stones and gems.

## 11.2 BEZOARS: PHILOLOGICAL, CONCEPTUAL, AND SYMBOLIC FRAMEWORK

Bezoar is a word of Persian origin. According to Ahmad al-Tīfāšī, an Arabic author born in Tīfas, a city in South Tunisia, in 1184, *Bāzahr* is a word of Persian origin, composed of *bāk* "purification," and *zahr* "poison," meaning "the one that purifies." The most important property of bezoar is to absorb poisons and bites of poisonous animals (Zilio-Grandi, 1999).

In the 18th century, the French chemist and pharmacist Pierre Pomet (1648−1715) who authored an influential compendium of drugs, published a drawing of a bezoar goat above a bezoar, displaying the typical concentric layers around the central foreign body (Pomet, 1712) (Fig. 11.1). The bezoar goat is the *capra aegagrus* from Europe, Central Asia, and Near East. Bezoars from other animals were also used as antidotes, especially in the 16th century. Other endogenous animal concretions and minerals have been used throughout history.

*Toxicology in the Middle Ages and Renaissance.* DOI: http://dx.doi.org/10.1016/B978-0-12-809554-6.00011-1

*Figure 11.1 Bezoar's goat and diagram. Drawing in Pierre Pomet,* Histoire générale des drogues, *Paris: Estienne Ducaston, 1644.* Sourced by Google Books.

In 1831, bezoars were classified according to their chemical composition. The concept of bezoar incorporated concretions found in the pineal gland, salivary ducts, stomach, intestines, gall bladder, and gall ducts of animals. An item pointed by Louden matches Western and Eastern bezoars, the two main groups relevant for the history of medicine:

> *Resinous, commonly called Oriental bezoar procured in Mallacca and sold for ten times their weight of gold. The Bezoar Histricis, Bezoar porcinum, pietro porco, or lapis malacencis, found in the gall bladder of the Indian porcupine were of this kind. They had a bitter resinous flavor. Diluted in water, were taken as aperients and stomachic. Some specimens of the Occidental bezoars, taken from the stomach or intestines of the goat or stag from were also of this kind, but they were sometimes composed of triple phosphate. Oriental and Occidental bezoars occasionally consisted of resin, and inert vegetable matter. (Louden, 1831)*

This classification by Louden concerns the significance of bezoars in the history of Western medicine until the beginning of the 19th century, when the medicinal use of bezoars, like theriacs and Hippocratic therapies in general, came to an end. More effective physiopathological approaches evolved, bringing a deep change to medical thinking. As the true science of poisons and antidotes took hold, bezoars, often richly mounted, were consigned to art and medical museums. Bezoars are simply hard masses of indigestible material such as plant fibers, seeds, hair, or chewing gum, found in the human stomach, and in the stomach and intestine of animals, especially ruminants, and other mammals.

## 11.3 BEZOARS AND THEIR GLORY

The glorious age of bezoars as fabulous medicines and apotropaic devices belongs to the past. Bezoars were highly prized in toxicology, pharmacy, and decorative arts. The Portuguese historian Jorge Manuel dos Santos Alves compared them to pearls, also a product of an organic reaction to foreign bodies. In his words, rarely had such a product persisted for so long worldwide in everyday life and in the human imagination. The Chinese highly esteemed bezoars since ancient times; they appear, for example, in the legends of the Zhou dynasty (770—256 BC). Their fame spread throughout Asia, largely due to their purported antidotal properties (Alves, 2005).

The rarity of bezoars, produced by a pathological process in only a few animals, was an important factor in their legendary status. For Ahmad al-Tīfāšī, who was acquainted with Chinese bezoars, the bezoar was an utmost precious and noble stone and the most powerful antidote against poisons. A commonly recommended procedure was to take scrapings, dilute them in olive oil or water, and ingest the solution. It could also be applied externally to wounds caused by the bites of venomous animals (Zilio-Grandi, 1999).

One of the most remarkable Arabic authors, the mathematician, geographer, astronomer, and encyclopedist, Abu al-RayanMuhamed ibnAhmad Al-Beruni (973—after 1050) wrote of the extent to which bezoars were appreciated in royal treasuries, noting their preciousness:

*It is collected in the treasuries of kings. Its price is very high and people are excessively covetous of it. I swear upon my life that it is the most precious of treasures as the soul enjoys it and benefits from it, more than from other jewels. (Said, 1989)*

In symbolic terms, counteracting the effects of poisons bore some relation to spiritual and mental elevation. Bezoars provided a healthy balance of physical and spiritual well-being. In primitive shamanic traditions, special qualities are attributed to rare objects. This concept also underlies ancient Chinese and Indian Ayurvedic medicine, which seeks harmony between men and gods, lower and upper levels of body and spirit, and ways to restore balance between the opposites to fight against disease. In traditional Chinese medicine, hundreds of animal parts are still dried and used as medicines nowadays. Bezoars are still amongst the most prized (Müller-Ebeling and Rätsch, 2011).

## 11.4 BEZOARS IN THE MIST OF HISTORY

In his chapter on bezoars, Ahmad al-Tīfāšī praised the surprising properties of stones found in the entrails of animals (Zilio-Grandi, 1999). These stones do not figure in the first Greek treatise, *On Stones*, by Theophrastus of Eresus (c. 371–287 BC), successor of Aristotle (384–22 BC) in the Peripatetic School. Theophrastus' work is free of magic and fables, except for the legend on the origin of Lyngurium (amber), as the congealed urine of the lynx (Caley and Richards, 1956). This book is regarded as the foundation of modern minerology and geology.

After Theophrastus' death, the rational and scientific perspective regarding the natural world declined. In the field of stones and minerals, the belief in magical properties of stones, which is as old as humanity, resurfaced. Hellenistic and Roman apocryphal lapidaries abounded, conveying magical, and astrological thought of Eastern origin (Halleux and Schamp, 1985).

In the 3rd century BC, Alexandrian medicine had been introduced into Mesopotamia. Syria had acquired the main body of Hippocratic teaching but retained many of the astrological features of the Assyrian–Babylonian Medicine. This dual system was studied by Syrian physicians for over thousand years (Campbell, 2006).

*Alexander's Romance* of unknown author, based on the legend of Alexander the Great (356–23 BC) inspired letters between Aristotle with his pupil Alexander on several matters, including science. Apocryphal lapidaries, i.e., books on stones, attributed to Aristotle are part of the hermetic literature, where the properties of stones are revealed like secrets.

While no genuine work of Aristotle on vegetables and minerals to accompany his celebrated History of Animals has ever been found, Arab documents readily filled the gap (Thorndike, 1923). In these works, legends on stones taken from animals abound. These stones concern real endogenous animal concretions, stones taken from birds' gizzards or stones of fabulous animal origin like Theophrastus' lynx stone. The Roman encylycopedist, Pliny the Elder (23–79 A.D.), cites these works frequently in the last volume of his work, dedicated to minerals. They are not, though, usually indicated as antidotes.

The Latin pseudo-Aristotelian *Secretum secretorum* or *Secreta secretorum*, a translation of the Arabic *Kitāb Sir al-asrār* (*The Book of the Secret of* Secrets) a work supposedly written from Aristotle to Alexander the Great, is found in about 500 manuscripts, dating from the 12th century onward. The Arabic work is extant in some 50 manuscripts. The earliest is a fragment dating from A.D. 941/330 (Manzalaoui, 1977). Stones, including those of animal origin, are regarded as part of the marvelous mysteries:

> *Full grete and marvelous is bothe in plants and in stones, bur fro mankind they ben hid. In primis, thefor, O Alexandre, I wolle yeve the amonge the secrets the greates Pat Purgh Goddes mytht shall helpthe to bring about thy purpose, and to kepe secré the priveté. Therfore take the stone animal, vegetable, and mynerall, the which is no stone, neither hath the nature of a stone (Manzalaoui, 1977).*

A manuscript from the 9th century *Kitāb Ğiranis*, an Arabian translation from an apocryphal Greek text whose material has come from the Hellenistic time, attributed to Hermes Trimesgistos, based on sympathetic magic, is divided in four chapters, each describing a plant, a stone, a bird, and a fish (Toral-Niehoff, 1997). Animal stones, taken from birds and from fish's heads abound. Some are of medicinal value used to treat leprosy, epilepsy, eye opacity, impotence, infertility, bladder stones, and sleep disturbances. None is indicated against poisons.

Saint Albertus Magnus or Saint Albert the Great (1193/1206−80), a catholic saint, a German friar and a bishop, considered the greatest German philosopher and theologian of the Middle Ages, conveys both the factual and magic traditions of stones in his book *Liber Secretorum Alberti Magni*, published in 1502. Among 45 entries, there are five stones associated with animals, four of these of medicinal use, and the following two useful against poisons. The first is draconites, an ammonite fossil, or snakestone:

> *Take the stone which is called Draconites, from the dragon's head. And if the stone be drawn from him alive, it is good against all poisons, and he that beareth it on his left arm, shall overcome all his enemies (Best and Brightman, 1999).*

This provenance described is fanciful, as well as its indication. The draconites is not to be taken orally but to be carried as an amulet to protect against poisonings and from all enemies. Serpent stones carry a long story in ancient and folk medicine. Medicinal use of

Serpent Stones first appears in Assyrian medical texts on cuneiform tablets from Ashurbanipal's library in the 17th century BC (Pymm, 2016, in press).

The other stone used against poisons is the *aetitis*, or eagle stone, because it was said to be found in eagles' nests. It was a geode, a hollow concretion containing crystals, a pebble or earthy matter (Best/Brightmon 1999). Saint Albert also followed the ancient medical tradition of the Near East. The last indication of the stone is against poison.

*As the men of Chaldea say, if poison be in thy meat, if the aforesaid stone be put in, it letteth that the meat out, the meat may be swallowed down. And if it be taken out, the meat is soon swallowed down, and I did see that this last was examined sensibly of one of our brethren. (Best and Brightman, 1999)*

Eagle stones were also used as amulets, mounted in silver (Duffin, 2012). The use of this stone against poisoned food was described by Ahmad al-Tīfāšī. Bezoars, it was thought, should be, mounted in rings and sucked to protect from a poisonous drink or animal (Zilio-Grandi, 1999). In the 16th century, bezoars were also carried as pendants as a protection against all evils and poisons (Fig. 11.2).

From the 8th century onward, bezoars began to figure in Arabic medical literature. Yuhannā Māsawayh (777−857), in Latin, Mesuë, Mesuë Senior, also known as Janus Damascenus or Serapion, featured bezoar as an alexipharmic, used in Syria and India against lethal poisonings and dangerous scorpion and snake bites (Serapiones, 1552).

*Figure 11.2 Oriental bezoar mounted on Indo-Portuguese golden filigree pendant. 16th century. Height 9.4 cm. Tavora Sequeira Pinto Collection (Oporto).*

From the 12th century onward, Arabic medicine exerted a growing influence on Western medicine. The great body of Greek medical works was read in Arabic translations. Until the 17th century, the works of Rhazes, Avicenna, Albucassis, Avenzoar, and Averroës received more attention than Hippocrates and Galen. The works of Serapion or Mesuë laid the foundations of a new pharmacy (Campbell, 2006). Despite this influence, bezoars were only recorded by some a few European authors until the sixteenth century.

Alfonso of Castile (1221–84), or King Alfonso X, the Wise, created a great translation center at Toledo, bringing Western scholars into more intimate contact with Arabian learning (Campbell, 2006). Alfonso of Castille authored a lapidary of which four manuscripts, housed in the Escorial Library, are extant. Animal stones are mentioned, such as a stone from the bull gall bladder, indicated in the treatment of skin, eyes, and ear diseases (Montalvo, 1981). This stone, later known as cow bezoar, is still used in Chinese traditional medicine (Fig. 11.3).

Bezoar was in Book I *"Della Piedra a que llamam Bezahar"* (*On Stone known as Bezoar*), as being of Indian origin to the astrological 11 Gemini, as an antidote and also useful against melancholia (Montalvo, 1981). This treatise provides the earliest reference to the bezoar stone in Galician–Portuguese, the ancestor of Portuguese language.

The Italian philosopher, physician, and astrologer Petrus Abanus (Peter of Abano or Pietro d'Abano) (1250–1316) listed the known poisons, symptoms of poisoning, and treatment. He described bezoar and its use as an antidote, according to Arabic sources (Abano, 1565).

*Figure 11.3 Chinese cow bezoar (author's collection). Photo by Ivo Miguel Barroso.*

Other European medical authors, who were widely acquainted with Arabic medicine, do not seem to have ever come across bezoars. Peter of Spain (born probably in 1205–77), the Portuguese pope and doctor, in his work *Thesaurum Pauperum* (*The treasury of the Poor*), written in 1272, mentioned the use of animal stones, such as stones from the brain of an eagle, the *Salpuga*'s jaw (*salpuga* is the Latin name of an animal unknown to us), deer's stomach or doe's stomach or vulva, but never mentioned bezoars. Constantin the African (1017–87), Tunisian doctor and professor of medicine at Salerno, later Benedictine monk who translated great masters of Arabic medicine, does not mention bezoars. Peter of Spain borrowed some recipes from him in which animal stones are used to treat infertility:

> XLV. For the woman to conceive 19. Item the stone found in the deer's stomach makes the woman carrying or taking it, conceive. Constantine. (Baptista, 2011)

The stone found in the deer's belly is surely a bezoar, but it is not identified as such. Its indication is also not of an antidote, as usually for bezoars.

## 11.5 CONCLUSION

Bezoar stones, endogenous animal concretions, have been widely touted particularly in the East and in the Middle Ages, as antidotes to poisoning. In the mist of their history, stories of bezoars were packaged with those of other endogenous, exogenous, and fabulous animal stones. While not absent from the Western canon, the medicinal and antidotal properties of bezoars were generally unknown in Europe.

## ACKNOWLEDGMENTS

I wish to thank Dr. Philip Wexler for the critical reading and revision of the manuscript, and Dr. Alvaro Sequeira Pinto for the kind permission to reproduce the image of the bezoar of his collection.

## REFERENCES

Abano, P, De venenis eorumque remediis, s/l, 1565, Omnia opera Guilielmi Gratoroli ex manu scripti exemplaribus collata, aucta atque illustrata.

Alves JMG: Realidade e mito em torno de um antídoto (séculos XVI e XVII). In Guillot C, Ptak R, Alves JMS, editors: *Mirabilia Asiatica. Produtos raros no comércio marítimo/Produits rares dans le commerce maritime/Seltene Waren im Seehandel*, Lisboa, 2005, Harrassowitz Verlag: Wiesbaden/Fundação Oriente, pp. 123–134.

Baptista, AM (coord.), Pereira, M.H.R. (trad. latim/port.), Reis, I./Robson (trad. port./ingl.), Petrus Hispanus. Thesaurus Pauperum, Tesouro dos Pobres, Treasury of the Poor, Lisboa, 2011, F. O., Hearbrain Consultores em Comunicação.

Best MR, Brightman FH, editors: *The book of secrets of Albertus Magnus, of the virtues of herbs, stones, and certain beasts, also a book of the marvels of the world (After the English edition of 1550)*, Boston, 1999, Weiser Books.

Caley ER, Richards JFC, editors: *Theophrastos on stones*, Ohio, 1956, Columbia University.

Campbell D: *Arabian medicine and its influence on the Middle Ages. Hertford*, 2006, Martino Publishing.

Duffin JC: A survey of birds and fabulous stones, *Folklore* 123(2):179–197, 2012.

Halleux H, Schamp J: *Les Lapidaires Grecs*, Paris, 1985, Les Belles Lettres, 1985.

Louden JC: *A magazine of natural history and journal of zoology, botany, minerology, geology and methereology*, IV, London, 1831, Longman.

Manzalaoui MA, editor: Secretum secretorum *nine English versions*, Oxford, 1977, University Press.

MONTALVO, S (ed. intr. trad. notas), Alfonso X, Lapidario Según el manuscrito escurialense H.I.15, Madrid, 1981, Gredos.

Müller-Ebeling, C & Rätsch, C, Tiere der Schamanen, Krafttier, Totem und Tierverbündete, Araau und München, 2011, A T Verlag.

Pomet, P, A complete history of druggs, London, 1712, J. and J. Bonnick, R. Wilkin and E. Wickseed.

Pymm R: 'Serpent stones' myth and medical application. In Duffin CJ, Gardner-Thorpe C, Moody RTJ, editors: Geology and medicine: historical connections. *Geological Society of London Special Publication*, 2016, in press.

Said, HM (org., transl., notes) (1989). Al-Beruni, English Translation of Al-Beruni's Book on Mineralogy: Kitab Al-Jamahir Fi Marifat Al-Jawahir, Islamabad, Pakistan, 1989, Hijra Council.

Serapiones, I De simplicium medicamentorum historia libri septem, extratum, ac Graecorum, præfertim Pauli Aeginetæ, Dioscorides, & Galeni commentarii quam accuratissime excerpti, Interprete Nicolo Mutono. Andrea Arrinabenium, Venetiis, 1552.

Thorndike L: *A history of magic and experimental science during the first thirteen centuries of our era*, II Vol, New York, 1923, Columbia University Press.

Toral-Niehoff, I (ed., transl. and notes), Kitāb Ğiranis, Die Arabische Übersetzung des ersten Kyranis des Hermes Trismegistos und die greichischen Parallelen, übersetzt und kommentiet. München, 1997, Herbert Utz Verlag.

Zilio-Grandi I: *Ahmad al-Tīfāšī. Il libro delle pietre preziose*, Venezia, 1999, Marsilio Editori.

# CHAPTER 12

# Fossil Sharks' Teeth as Alexipharmics

Christopher J. Duffin
The Natural History Museum, London, United Kingdom

## 12.1 FOSSIL SHARKS' TEETH OR GLOSSOPETRAE

A large oil-painted wooden panel in the Robert Lehman Collection (New York Metropolitan Museum of Art) depicts the inside of a goldsmith's shop. The bottom edge of the panel carries an inscription which indicates that "Petrus Christus made me in the year 1449." Christus (c. 1410−75/1476) was an Early Netherlandish or Flemish Primitive painter who came to prominence following the death of Jan Van Eyck in 1441. His study shows an upper class, young betrothed couple visiting the shop to seek purchases for their forthcoming wedding. The premises are replete with the varied wares of the goldsmith (believed by many to be a representation of St Eligius or Saint Eloi), including a number of items that were used as contemporary prophylactics and defenses against poisoning (Duffin, 2011). On the back wall of the shop, hanging on a small peg between a card of brooches and a rock crystal vessel, just above a branching piece of pink coral resting on the shelf beneath, is a pair of grey, triangular fossil sharks' teeth, each set in a gold base and suspended from a gold ring by means of fine gold chains. These are "glossopetrae" or "tonguestones" which, at the time of execution of the painting, had enjoyed over 200 years of use by both religious and secular European nobility as agencies to both detect poisons and counteract their effects.

Glossopetrae were first referred to by Pliny the Elder (AD 23−79) in his encyclopedic compendium of Roman folklore and only surviving work, *Historia Naturalis* (Natural History), probably written around AD 73. Pliny indicates, somewhat critically, contemporary belief in the

Toxicology in the Middle Ages and Renaissance. DOI: http://dx.doi.org/10.1016/B978-0-12-809554-6.00012-3

supposed supernatural origins of the stones and their resulting magical properties as follows:

> *Glossopetra, which resembles the human tongue, is not engendered, it is said, in the earth, but falls from the heavens during the moon's eclipse; it is considered highly necessary for the purposes of selenomancy [divination by the appearance and phases of the moon]. To render all this however, still more incredible, we have the evident untruthfulness of one assertion made about it, that it has the property of silencing the winds*
>
> **Bostock and Riley (1857:449-50)**

This information was effectively recycled, firstly by the 3rd century Latin grammarian and compiler Solinus (Polyhistor, 37:19; Golding, 1587, unpaginated), and then by Isidore of Seville (Isidorus Hispalensis, c. 560–c. 636) in his highly influential digest of universal knowledge, the *Etymologiarum* (XVI. 15.17; Throop, 2005 unpaginated). Then follows something of a hiatus in the literature of glossopetrae; they reappear with a virtually fully formed associated folklore, ascribing them alexipharmic properties (as antidotes to poisons), in western lapidary literature (books about stones) from around the late 13th and early 14th centuries onward. Their records in contemporary inventories point to a well-established material culture pervading the higher echelons of society.

Although absent from Marbode's Lapidary (11th century) and its various derivatives, glossopetrae are seemingly first cited in the Lapidary of Jean Mandeville (14th century), who explains that, in addition to inculcating honest and graceful speech in the bearer, they resist poisons and change color in their presence (del Sotto, 1862:101). The Peterborough Lapidary (15th century) reports that glossopetrae "defendeth a man from venome," but adds the rider that "yf a mane be poysened it kepeth him" (Evans and Serjeantson, 1933:98). The slightly later Sloane Lapidary (16th century) reiterates advice from the Peterborough Lapidary, commending that these "Tongs of Adders ... should be set in silver, both for kings [&] lords at their meate, so yt they mey be kept ye safer from poison" (Evans and Serjeantson, 1933:130). In this instance, the stone is indicated as breaking into a sweat in the presence of poison.

The means of action of this stone clearly rely on sympathetic magic. The V-shaped root of the tooth evokes the forked tongue of a snake, many species of which are venomous to some degree; the morphology of the tooth was a "signature" indicating the broad uses to which it could be put using the principle of "like cures like" (*similia similibus curantur*). Just as various parts of snake anatomy were components of alexipharmic preparations such as Mithridatium, the snake-like tonguestones, were seen as powerful antivenins which could also confer higher qualities of speech to the bearer. Colloquial terms, particularly in Germany, also associated these tonguestones, on the basis of their morphology, with otters, birds (especially woodpeckers) and mankind (see below).

Jean de Mandeville's Lapidary indicates that the tonguestone was carried on the person (del Sotto, 1862:101). The wearing of Recent and fossil shark's teeth as amulets has a long history, and persists in some regions today. Shark's teeth mounted in exquisitely worked gold housings form part of the grave goods found at various Etruscan sites (4th to 5th centuries BC), especially in Italy (e.g., Cherici, 1999). Fossil shark tooth amulets are reasonably well known in European museum collections (e.g., Hansmann and Kriss-Rettenbeck, 1977: 171; Oakley, 1985 pls I, II). The specimen in Fig. 12.1, for example, has the root, which is the longest part of the tooth, encased in a silver mount,

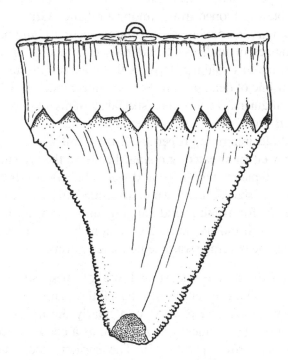

*Figure 12.1 Fossil shark tooth (glossopetra) pendant (British Museum, OA. 1386). The tooth belongs to* Otodus megalodon, *probably from the Miocene of Malta, and comes from a posterolateral position in the left upper dentition.* Original.

clearly designed to hang from a chain or other mount. The question might legitimately be raised as to whether mounted shark's teeth such as these were actually used to provide a prophylactic or protective function against poisons. This is certainly the case in some specimens contained in historical materia medica collections. There is , for example, a group of medicinal glossopetra amulets in the collection of John Burges (1745–1807), a passionate Georgian collector of materia medica artifacts whose collection is now housed in the Royal Pharmaceutical Society Museum (London) (Duffin, 2013).

## 12.2 TABLEWARE

It is eminently possible that some of the glossopetra pendant survivals formed part of a popular item of late medieval and renaissance tableware. High status Europeans lived in fear of poisoning; the preferred means of dispatch seemed often to involve arsenical compounds which were odorless, tasteless, and soluble. Protection could be achieved by serving food, and drinks in particular, in any number of a range of exotic containers, including goblets fashioned from, amongst others, "Griffin's claws," unicorn horn (of narwhal or rhinoceros origin), coconuts, and bezoar stones, which are masses of undigested material accumulating in the digestive tract (see Lavers, this volume; Barroso, this volume). Prominent amongst this particular line of defense were Natternzungenbaum ("Adder's tongue trees") or Languiers. These were specialist items of orfèvrerie (plate) commissioned from gold- and silversmiths. Their efficacy relied upon the supposed alexipharmic properties of the fossil shark's teeth, and the belief that they either sweated or changed color in the presence of poison, as recorded in lapidary literature, and were able effectively to neutralize it. The mounted shark's teeth were suspended from a central tree-like structure, ready for picking and dipping into the wine before it was drunk. If the tooth did not undergo a color change on being extracted from the wine, the beverage was deemed safe to drink.

An impression of the widespread use of these structures can be gained from surviving inventories. One of the main consumers of this item of tableware was the papacy, particularly during (but not limited to) the period of their residency at Avignon as a consequence of conflict with the French crown (1309–77). The earliest record in the papal inventories appears to be of "Rami sive arbores cum linguis serpentinis" ("Branches or trees with serpent's tongues") dated 1295 and belonging to Pope Boniface VIII (Pogatscher, 1898:173). Numerous further tree-like specimens, some bearing as many as 11 suspended shark's teeth on up to 8 branches of red coral, plus isolated "tongues," are listed in records dating right up to the late 16th century (Pogatscher, 1898).

Noblemen also treasured these poison detectors, often referred to as *probae* (Latin; *èpreuves* in French), or salts (*salières*). The inventories of Jean, first Duc de Berry (1340–1416), record at least 33 different serpent-tongue *èpreuves* (Guiffrey 1894, De Laborde 1872), whilst other records are scattered through the inventories of royalty and the ennobled, including such luminaries as Edward I of England (1239–1307), Charles V of France (1338–80), and the Dukes of Anjou and Normandy (Nichols, 1787:352; Labarte, 1879; De Laborde 1853). Seemingly, the earliest recorded example dates from 1266 and belonged to Odo of Burgundy, Count of Nevers (1231–66) (Schiedlausky, 1989:26).

Four remarkable surviving examples are known, three of which retain suspended shark's teeth (Duffin, 2012). A specimen in the Green Room of the Staatliche Kunstsammlungen in Dresden (IV 108) consists of a silver base with Jesse, the Father of David, flanked by a snake and reclining at the base of a tree. Six long pedicels then emerge through a canopy of serrated silver leaves, each terminating in a drooping flower from which a tooth of *Isurus* is suspended. In the crown of the tree, Mary, with the baby Jesus in her lap, leans against a large specimen of *Otodus megalodon*.

A 32-cm-high coral tree with 14 *megalodon* teeth hanging from the branches, all surmounting a silver gilt base, is found in the treasury of the German Order in Vienna (K0-37). A second coral tree, but lacking the suspended shark's teeth, is recorded from the Treasury of the Cathedral Museum at Mdina in Malta. The final survivor dates from the 15th century are kept in the Kunsthistorisches Museum of Vienna (KK 99). A long stem extends from the petaloid base, leading to a crown of silver flowers surmounted by a goblet ornamented with topaz. Fossil shark's teeth project radially from the central stand.

## 12.3 PROVENANCE OF THE TEETH

Whilst teeth from a wide range of fossil shark genera, and collected from any suitable rock could be useful in detecting poison, Miocene specimens of *O. megalodon* from Malta were praised universally as the most efficacious, based partly on a folkloristic association with St Paul. Shipwrecked on Malta during his journey to trial at Rome, the incident of the adder in the woodpile (Acts 28:1–6) supposedly caused St Paul to rid the island of all poisonous serpents. The power of his preaching meant that his tongue was able to penetrate solid rock, leaving a representation of itself behind. These "Ilsien San Pawl" or St Paul's Tongues are actually fossilized lamnid shark's teeth (Zammit-Maempel, 1989:18). Other representations of different parts of St Paul's anatomy, and powdered rock from the cave which he occupied at Rabat (St Paul's Earth or *Terra Sigillata Melitensis*) are also credited with therapeutic and alexipharmic properties.

A brisk trade in fossil sharks' teeth grew up between Malta and European apothecary shops. This was especially important during the middle part of the 16th century when handbills promoting the therapeutic qualities of St Paul's Earth, Ilsien San Pawl and *Occhi di Serpi* (Snake's eyes—fossilized fish teeth) were published in Latin, French, Italian, and German (Zammit-Maempel, 1978). One example is illustrated in Fig. 12.2. It was recommended that the teeth be suspended either from the neck or the arms and that, if they should fail to protect adequately against poison, they could nevertheless be relied upon to effect a cure.

The notices often close with dire warnings against purchasing ineffectual fake tongues. In 1565, Conrad Gessner devised a test by which to distinguish genuine glossopetrae from the "teeth of the still existing monsters" (i.e., modern Great White Sharks) which were potential substitutes. A fine thread should be very carefully wound around the stone so that none of its surface was eventually visible, but the string circuits never overlapped each other. The object was then laid in hoar frost. The genuine article could be distinguished from the fake by the thread becoming damp (Gessner, 1565 folio 62). Bartholinus (1643:200) notes that even "calcareous concretions" were substituted for the genuine Ilsien San Pawl.

It was not until the second half of the 16th century that Conrad Gessner was able to suggest that tonguestones were the petrified remains of the teeth of ancient sharks. His assertions were the subject of considerable doubt, and the idea was not accepted until over 100 years later, when the true nature of fossils was beginning to emerge onto a scientific footing.

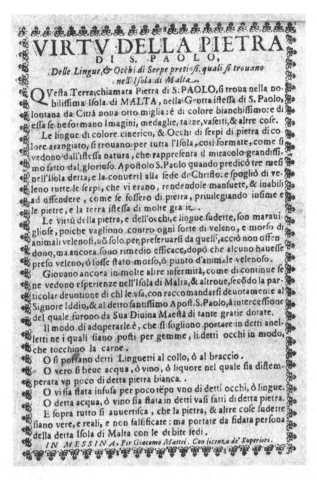

*Figure 12.2 Handbill in Italian promoting fossil sharks' teeth.* Coll. BNU de Strasbourg, R.5 Fol. 4.

## 12.4 LAPIS DE GOA

Fossil shark's teeth were also counted amongst the ingredients used in the manufacture of the popular Lapis de Goa, or Goa Stone. A confection invented by a Portuguese Jesuit lay-brother, Gaspar Antonio, probably during the mid-16th century, the Lapis de Goa was produced in response to the contemporary shortage of bezoar stones. One recipe preserved in the Vatican Archives records that "Lingui di San Pauli" were pulverized and combined into a paste, together with a variety of precious stones, coral, ambergris, musk, and other ingredients (Amaro, 1988/1989). The mass was fashioned into balls, dried, and then covered with gold leaf before being exported from the missionary

*Figure 12.3 Goa Stone and holder.* Wellcome Library, London.

station in Goa to European apothecary shops. Many of the individual ingredients of the artificial stone were credited independently with anti-toxic properties and, combined together, were seen as working synergistically as a cordial and alexipharmic medicine (Duffin, 2010a,b). The gold-plated ball was highly revered and, like Bezoar Stones, was often housed in an elaborate filigreed silver or gold cage-like container (Fig. 12.3). Scrapings from the stone were often administered in a suitable liquid medium, often Sack, wine, or beer, or could be taken as a snuff. It acted against any poisons in the body, whether exogenous or related to humoral imbalance, by a sudorific action (promoting sweat). Supposedly effective against a wide range of conditions, it was even used (unsuccessfully!) to treat royalty (see details in Duffin, 2010a,b). The Goa Stone was popular as a medicine right through to the mid-18th century, and several specimens are extant in modern materia medica collections.

## REFERENCES

Amaro AM: A famosa *Pedra Cordial* de Goa ou de Gaspar António, *Rev de Cultura* 88-XX:1988/1989.

Bartholinus T: *Historiarum anatomicarum & medicarum rariorum centuria V. & VI.* Hafniae, 1643, Henrici Gödiani.

Bostock J, Riley HT: *The Natural History of Pliny*, Volume VI, London, 1857, Henry G. Bohn.

Cherici A: Amuleti nei corredi funebri Paleoveneti e dell'Italia antica. V. In Protostoria de Venetorum Angulus. *Atti del XX convegno di Studi Etruschi ed Italici, Portogruaro, Quartod'altino, Este, Adria. 16−19 ottobre 1996 Pisa-Roma*, 169−216, 1999.

De Laborde M: *Notice des émaux, bijoux et objets divers, exposés dans les galleries du Musée du Louvre*, 2, Paris, 1853, Vinchon.

De Laborde M: *Glossaire Français du Moyen Age*. Paris, 1872, Adolphe Labitte.

Duffin CJ: Lapis de Goa: the "Cordial Stone", *Pharm Hist (Lond)* 40(2):22−32, 2010a.

Duffin CJ: Lapis de Goa: the "Cordial Stone"—part two, *Pharm Hist (Lond)* 40(3):42−46, 2010b.

Duffin CJ: Glimpse into a 15th century Goldsmith's Shop, *Jewellery History Today Issue* 11:3−4, 2011.

Duffin CJ: Natternzungen-credenz: tableware for the Renaissance nobility, *Jewellery History Today, Issue* 14:3−5, 2012.

Duffin CJ: Porcupine stones, *Pharm Hist (Lond)* 43(1):13−22, 2013.

Evans J, Serjeantson MS: *English medieval lapidaries*, London, 1933, Oxford University Press for the Early English Text Society.

Gessner C: *De Rerum Fossilium Lapidum et Gemmarum maximè, figures & similitudinibus Liber: non solùm Medicis, sed omnibus rerum Naturae ac Philologiae studiosis, utilis & iucundus futurus*. Tiguri, 1565, Jacobus Gesnerus.

Golding A: *The excellent and Pleasant Worke of Iulius Solinus Polyhistor. Contayning the noble actions of humaine creatures, the secretes & prouidence of nature, the description of Countries, the maners of the people: with many meruailous things and strange antiquities, seruing for the benefit and recreation of all sort of persons. Translated out of Latin into English by Arth. Golding, Gent.* London, 1587, I. Charlewoode.

Guiffrey J: *Inventaires de Jean Duc de Berry*, Paris, 1894, Ernest Leroux.

Hansmann L, Kriss-Rettenbeck L: *Amulett und Talisman. Erscheinungsform und Geschichte*, Munchen, 1977, Verlag Georg D.W. Callwey.

Labarte J: *Inventaire de Mobilier de Charles V, Roi de France*, Paris, 1879, Imprimerie National.

Nichols J: *Liber quotidianus contrarotulatoris garderobae. Anno regni regis Edwardi primi vicesimo octavo. A.D. MCCXCIX & MCCC. Ex codice MS. in biblioteca sua asservato typis edidit Soc. Antiq. Londinensis.* London, 1787, Excudebat J. Nichols: Prostant venales apud B. White et fil. Jacob. Robson et Gul. Clarke, Georg. Leigh et Joan. Sotheby, Gul. Browne, necnon T. et J. Egerton.

Oakley KP: Decorative and symbolic uses of vertebrate fossils. *Pitt Pivers Museum, University of Oxford, Occasional Papers on Technology*, vol 12, 1985, pp 1−60.

Pogatscher H: Von Schlangenhörnern und Schlangenzungen vornehmlich im 14. Jahrhunderte (Mit Urkunden und Akten aus dem vaticanischen Archive), *Röm Quart Christ Altertum Kirch* 12:162−215, 1898.

Schiedlausky G: Natterzungen: Ein Leitfossil in der Geschichte mittelalterliche Giftfurcht, *Kunst und Antiquitaten* (1989, no. 6):25−31, 1989.

del Sotto I.: *Le Lapidaire du Quatorzième Siècle*. Vienna, 1862, Impr. impériale et royale de la cour et de l'état, et de l'état, [Slatkine reprint, Geneva, 1974].

Throop P: *Isidore of Seville's Etymologies. The complete English translation of Isidori Hispalensis Episcopi* Etymologiarum sive Originum Libri XX, 2, Charlotte, 2005, Lulu.com.

Zammit-Maempel G: Handbills extolling the virtues of fossil shark's teeth, *Melita Hist* 7 (3):211−224, 1978.

Zammit-Maempel G: The folklore of Maltese fossils, *Pap Mediterr Soc Stud* 1:1−29, 1989.

# CHAPTER 13

# Catherine La Voisin: Poisons and Magic at the Royal Court of Louis XIV

**Benedetta F. Duramy**
Golden Gate University School of Law, San Francisco, CA, USA

## 13.1 INTRODUCTION

In the decade between the late 1660s and the early 1670s, influential members of the French nobility were mysteriously poisoned to death. Fearing for his own safety and that of the royal family, Louis XIV appointed Nicolas de La Reynie, the Lieutenant General of the Paris Police, to prompt an investigation, which unveiled the diffuse use of the practice of poisoning and led to the "Affair of the Poisons" scandal (Benedetta Faedi Duramy, 2012c).[1]. The first trial was that of the serial poisoner Marie-Madeleine Marguerite d'Aubray, Marquise de Brinvilliers, who, before being tortured with water, confessed of having poisoned her father and her two brothers as well as having attempted to poison her sister-in-law and her husband several times (Benedetta Faedi Duramy, 2012a).[2]. Just before her execution, she revealed that "[o]ut of so many guilty people must I be the only one to be put to death? ... [And yet] half the people in town are involved in this sort of thing, and I could ruin them if I were to talk" (Somerset, 2003). Frightened by such allegations, the Court of Paris was eager to discover the name of her accomplices and purveyors as well as the secret of the poisons and antidotes she had used. In fact, President Lamoignon of the Paris Parliament declared, "it is in the public interest ... that Mme de Brinvilliers's crimes end with her, and that she makes a declaration that will help us to prevent the continued use of poison" (Mossiker, 1969). However, the Marquise de Brinvilliers did not betray her accomplices

---

[1]Excerpts of this chapter have previously appeared in Benedetta Faedi Duramy, Women and Poisons in 17th Century France, 87 Chicago-Kent L. Rev. 347 (2012).
[2]For a short biography of Marquise de Brinvilliers, see Benedetta Faedi Duramy, Brinvilliers, Marquise de, in 2 Women Criminals: An Encyclopedia of People and Issues, 333–335 (Vickie Jensen ed., ABC-CLIO, 2012).

Toxicology in the Middle Ages and Renaissance. DOI: http://dx.doi.org/10.1016/B978-0-12-809554-6.00013-5

but only admitted using arsenic, vitriol, and venom of toad as poisons and milk as antidote to kill her victims. Others, to whom she alluded, though, ended up being implicated in the Affair of the Poisons, and many were then prosecuted and sentenced to death. The Marquise de Brinvilliers' trial caused such a sensation that the penitentiaries of Notre Dame declared, "the majority of those who had confessed to them for some time accused themselves of poisoning somebody" (Frantz Funck-Brentano, 1901). To prosecute so many cases, Louis XIV established a special tribunal, known as the Chambre Ardente, in 1679, which ruled for over 3 years, issuing 319 subpoenas, arresting 194 individuals, and sentencing 36 of them to death. This chapter examines, in particular, the life and involvement in the scandal of the sorceress Catherine La Voisin, who, after being prosecuted, was burned alive for providing members of the royal court, including the Marquise de Montespan (Benedetta Faedi Duramy, 2012d)[3], one of the favorite mistresses of Louis XIV, with magic powders and venomous potions.

## 13.2 CATHERINE LA VOISIN

On March 12, 1679, La Reynie ordered to arrest the sorceress Catherine Deshayes while she was leaving the church of Notre Dame after hearing mass. Known as La Voisin (Benedetta Faedi Duramy, 2012b)[4], she was considered one of the greatest criminals at the time, belonging to the region of crime where "human life is publicly trafficked in...[.] death is almost the only remedy employed in family embarrassments; impieties, sacrileges, abominations are common practices..." (Frantz Funck-Brentano, 1901). Born in 1640 in France to a poor woman, who was a sorceress herself, Catherine Deshayes was initiated into chiromancy and face reading at the age of nine. She married Antoine Montvoisin, whose businesses in the silk trade and jewelry led him into bankruptcy. As a result, her husband lapsed into heavy drinking and to violently venting his frustrations out on her. Having to support him and her numerous children, she turned into the criminal, but lucrative, business of fortune telling, abortion, and the preparation of poisons (Mollenauer, 2007).

[3]For a short biography of Marquise de Montespan, see Benedetta Faedi Duramy, Montespan, Marquise de, in 2 Women Criminals: An Encyclopedia of People and Issues 542–544, (Vickie Jensen ed., ABC-CLIO, 2012).

[4]For a short biography of Catherine La Voisin, see 2 Benedetta Faedi Duramy, Catherine La Voisin, in 2 Women Criminals: An Encyclopedia of People and Issues 507-SOB, (Vickie Jensen ed., ABC-CLIO, 2012).

Once, when La Lepere, one of her friends, warned her that it was dangerous to engage in such crimes, she responded: "You are mad! The times are too bad. How am I feed my family? I have six persons on my hand!" (Frantz Funck-Brentano, 1901).

Her marriage was so unhappy that she tried multiple times to get rid of her abusive husband without success and engaged in many love affairs with other wizards and alchemists, including Andre' Guillaume, the executioner of Paris, who beheaded Madame de Brinvilliers, and the magician Lesage, who was later also implicated into the Affair of the Poisons (Somerset, 2003). When the latter advised her to get rid of her husband, "a sheep's heart was bought to which Lesage did something, and then it was buried in the garden behind the gate...[.] Monvoisin was seized with severe pain in the stomach. He cried out that if there was anybody who wished to do it for him, he had better shoot him at once instead of letting him linger. La Voisin, struck with remorse, hastened to the Augustines to confess and obtain a general absolution; she took the sacrament, and on her return compelled Lesage to undo his wicked charms" (Frantz Funck-Brentano, 1901).

During her interrogation with La Reynie, La Voisin herself alluded to her powers: "Some women asked if they would not soon become widows, because they wished to marry someone else; almost all asked this and came for no other reason. When those who come to have their hands read asked for anything else, they nevertheless always come to the point in time, and ask to be ridden of someone; and when I gave those who came to me for that purpose my usual answer, that those they wished to be rid of would die when it pleased God, they told me that I was not very clever" (Frantz Funck-Brentano, 1901). Her servant Margot testified, "La Voisin is to-day dragging a great ruck down with her—a long chain of persons of all sorts and conditions" (Frantz Funck-Brentano, 1901). And others recalled, "[a]t that time, La Voisin had as much money as she wanted. Every morning before she rose, people were waiting for her, and she had visitors all the rest of the day: after that, in the evening, she kept open house, engaged fiddlers, and enjoyed herself thoroughly; this went on for several years" (Frantz Funck-Brentano, 1901).

A high priestess of Christian congregations in Paris and a pious worshipper, La Voisin conceived her occult powers as a gift from God. Proud of her trade as sorceress, she "did not stick at any expense that seemed likely to augment her glory. She delivered her oracular sayings clothed in a robe and a cloak specially woven for her, for which she paid 15,000 livres. The queen herself had no finery more beautiful than this "imperial robe," which "was the talk of all Paris." The cloak was of crimson velvet studded with 205 two-headed eagles of fine gold, lined with costly fur; the skirt was of bottle-green velvet, edged with French point. Even her shoes were embroidered with golden two-headed eagles. The mere weaving of the eagles on the cloak cost 400 livres" (Frantz Funck-Brentano, 1901). Her clients, who primarily belonged to the highest rank of the French society, were likely reassured by La Voisin's religious devotion to her magical practices. One of her contemporaries, the Marquis de la Reviere, noted that La Voisin "was full of delicious little secrets for the ladies ... for which the gentlemen could be grateful .... [She] could make a lady's bosom more bountiful or her mouth more diminutive, and she knew just what to do for a nice girl who had gotten herself into trouble" (Mossiker, 1969). Madame Montvoisin received her clients in a small room hidden at the back of the garden of her house located in Ville-Neuve, a secluded area in the northern outskirts of Paris (Mollenauer, 2007).

Upon her arrest, the search of her premises revealed all sorts of magic powders, venomous potions, sacrilegious objects, "grimoires or black books (primers for Satanists and necromancers, the ABC's of Abracadabra), sacerdotal vestments and paraphernalia, a cross, incense, black tapers; a mysterious oven in a garden pavilion, redolent of evil, noxious fumes; fragments of human infants' bones in the ashes," (Mossiker, 1969) as well as a long list of her clients. She was indeed accused of having attempted to poison her husband several times at the instigation of her paramour, Lesage, and of having performed abortions for a fee, burying the premature infants in her garden. On March 17, 1679, when Lesage was arrested as well, he provided Nicolas de La Reynie with detailed accounts of La Voisin's business of abortions, traffic in poisons, and her customers. He revealed that a small oven was hidden in her house "where the bones were burned if the infant body seemed too large to layaway in a garden grave" (Mossiker, 1969).

Black masses and monstrous rites were also performed on women's bodies "laid stark naked, without even their chemise, upon a table which served as altar; their arms stretched out, and...[.] and a taper in each hand...[.] [t]he chalice was placed on the bare belly. At the moment of the *offertoire*, a child had its throat cut" (Frantz Funck-Brentano, 1901). After a long needle had been placed into the child's neck "the blood of the expiring victim was poured into the chalice, and mixed with the blood of bats and other materials obtained by filthy means. Flour was added to solidify the mess, which was thus made to resemble the Host, to be consecrated at the moment when, in the sacrifice of the mass, transubstantiation takes place" (Frantz Funck-Brentano, 1901). Such testimony finally solved the mysterious "great disturbance in Paris in 1676, when there were seditious gatherings and mobs and runnings to and from in several parts of the city, through the rumor that people carried off children to cut their throats, though no one then understood what the cause of the rumor could be" (Frantz Funck-Brentano, 1901) La Reynie wrote to Marquis of Louvois, the Secretary of State for War.

La Voisin denied everything, clarifying that the oven was used to bake her "petits pates" (little pastries) and that "the only drugs to be found in her house were purgatives for her personal use and that of her family" (Mossiker, 1969) or that they were used to cure pimples and remedy headache. Further accusations against Madame Montvoisin came from other prisoners claiming that her secret "to empty" pregnant clients consisted of injecting lethal liquid with a syringe:

- What's her secret to empty women or girls that are pregnant?
- Yes, it's basically water, and everything depends on the way the syringe is used.
- Until which stage of the pregnancy can she do it?
- Any time, especially when they are persons of quality, who must preserve their honor and don't want to make it public. As long as she can feel the baby moving before using her remedy, she will make the baby come out and baptize her/him. Then, she herself brings the baby in a box to the gravedigger, to whom she gives a coin of 30 cents, in order to bury her/him in a corner of the cemetery, without telling the priest or anyone else (Mongredien, 1953).

Lesage also revealed that most of La Voisin's visitors belonged to the King's entourage, and even a maid of the Marquise de Montespan, one of the favorite mistresses of Louis XIV, had purchased love powders from her. Catherine La Voisin initially counterclaimed that "nothing but beauty balms and skin lotions" were procured for her clients (Mossiker, 1969). In response to the allegations against Madame de Montespan, the King solicited La Reynie "to continue the questioning of certain of the prisoners ... [and] to proceed as speedily as possible with such interrogations, but to make the transcripts of these responses on separate folios, and to keep these folios apart from the official records of the rest of the investigation" (Mossiker, 1969).

Indeed, the interrogations of Lesage and other informants containing allegations against the Marquise de Montespan were all scrupulously removed from the trial dossier, never handed over to the judges of the Chamber for their scrutiny, but instead delivered exclusively into the trustworthy hands of La Reynie.

Eventually, Catherine La Voisin admitted that some of her customers were indeed prominent figures of the nobility, but firmly denied ever having served the Marquise de Montespan or even meeting with her. During her last interrogation, when she was subjected to the boot torture, she admitted, "Paris is full of this kind of thing and there is an infinite number of people engaged in this evil trade" (Somerset, 2003). However, regarding her customers, she only confessed, "a great number of persons of every sort of rank and condition addressed themselves to her to seek the death of or to find the means to kill many people" (Somerset, 2003), but refused to utter further names. In her last testimony, her confessor noted that she finally admitted, "I am loaded with so many crimes that I could not wish God to work a miracle to snatch me from the flames, because I cannot suffer too much for the sins I have committed" (Frantz Funck-Brentano, 1901). Catherine La Voisin was finally burned at the stake in 1680. Spectators at her execution reported that "five or six times, she pushed aside the straw, but finally the flames leaped up, enveloped her, and she was lost to sight. ... So, there you have the death of Mme Voisin, notorious for her crimes and impiety" (Mossiker, 1969).

## REFERENCES

Benedetta Faedi Duramy, Marquise De Brinvilliers, 2 women criminals: an encyclopedia of people and issues, 333–335 (Vickie Jensen ed., ABC-CLIO, 2012a).

Benedetta Faedi Duramy, Catherine La Voisin, 2 women criminals: an encyclopedia of people and issues 507-sob, (Vickie Jensen ed., ABC-CLIO, 2012b).

Benedetta Faedi Duramy, Women and poisons in 17th century France, 87 Chicago-Kent 1. Rev. 347 (2012c).

Frantz Funck-Brentano, Princes and poisoners: studies of the court of Louis XIV, 117, 121, 151, 149, 145, 146, 147–150, 155, 156, 159, (1901).

Lynn Wood Mollenauer, Strange revelations: magic, poison, and sacrilege In Louis XIV's France, 22, 21, (2007).

Georges Mongredien, Madame De Montespan Et L'affaire Des Poisons, 45–46, (1953).

Frances Mossiker, The affair of the poisons: Louis XIV, Madame De Montespan, and one of history's great unsolved mysteries, 146, 177, 179, 185, 186, 219, (1969).

Anne Somerset, The affair of the poisons: murder, infanticide and satanism at the court of Louis XIV, 40, 153, 231, (2003).

# CHAPTER 14

# A Late Medieval Criminal Prosecution for Poisoning: The Failed Murder Trial of Margarida de Portu (1396)

## Caley McCarthy[1] and Steven Bednarski[2]

[1]McGill University, Montreal, QC, Canada [2]St. Jerome's University in the University of Waterloo, Waterloo, ON, Canada

On October 16, 1394, the criminal court of Manosque, in southern France, initiated an inquest against Margarida de Portu, who allegedly killed her husband with magic or poison. Public rumor, inspired by the machinations of Margarida's brother-in-law, Raymon Gauterii, moved the criminal judge to investigate. The trial, which spanned 5 months and many folios, incorporated testimonies, including that of Vivas Josep, a Jewish physician whose expert testimony drew on learned toxicological theories to disprove the accusation of poisoning.

The accusations and consequent inquest against Margarida de Portu serve as a lens through which to view conceptualizations of poison in the late medieval Mediterranean. Although the trial's conflation of poison and sorcery are consonant with the popular and learned discourses on these crimes in the later Middle Ages, Vivas Josep's medical investigation applied a learned toxicology which sought to unveil the occult nature of poison that bound its action to magic in the medieval imagination.

The trial of Margarida de Portu, conserved in the Archives départementales des Bouches-du-Rhône (ADBDR 56H 1001 ff 32−60), has been the subject of two previous studies. In a 2002 article, Andrée Courtemanche used the case to shed light on the place of medical expertise in Manosquin judicial rituals. More recently, Margarida formed the subject of the microhistory, *A Poisoned Past. The Life and Times of Margarida de Portu, A 14th-Century Accused Poisoner* (Bednarski, 2014). We provide here a summary of both works to

Toxicology in the Middle Ages and Renaissance. DOI: http://dx.doi.org/10.1016/B978-0-12-809554-6.00014-7

illuminate popular and learned conceptualizations of poison contained in late medieval Provence.

At about the age of 16, Margarida de Portu emigrated from her native village of Beaumont to the nearby town of Manosque to marry Johan Damponcii in the spring of 1394. Manosque, situated between the religious and commercial centers of Avignon, Aix, Marseille, and Rome, had a population of approximately 5000 before the arrival of the Black Death in 1348, when, as expected, this number plummeted. There were artisans, craftspeople, and professionals, both Jewish and Christian. Although they occupied different quarters in the town, the adherents of these two faiths cohabited relatively harmoniously, mingling in the day-to-day affairs of business and pleasure. By Margarida's lifetime, the city fell under the governance of the Hospitallers (a medieval Catholic military order), though, in earlier years, the order shared this right with the counts of Forcalquier. In an act of bestowing his own poisoned gift, the last count relinquished his rights to Manosque in favor of the Hospitallers, but not before he largely emancipated his former subjects. Still, the Hospitallers assumed sole governance and, with it, the ancient customary rights to dispense all forms of justice, civil, and criminal. Their courts operated under Roman law and, in the mid-13th century, followed the typical shift from accusatorial to inquisitorial process. Formal denunciations, as well as public rumor and gossip, inspired judicial inquests, and it was the latter that brought Margarida before the court in 1394. The inquest clearly reflects the interests of her brother-in-law, Raymon Gauterii, who was a notary and thus familiar with the rules in Roman law that prevented a denouncer from testifying. He, therefore, chose instead to spread rumors about the suspicious circumstances under which his brother had died and waited for the *fama publica*, the public rumor, to move the judge to act ex officio and to investigate.

The trial begins by outlining Raymon's suspicions. He alleged that Margarida had "handed her husband, to whom she stands forth espoused through the matrimonial knot, on account of a false and harmful suggestion in her damnable soul, over to the defeat of death suddenly and, moreover, through the use of sorcery or venom, since venom is more painful to slay with than the sword" (ADBDR 1001 fo. 32v). In order to strengthen the plausibility of these accusations, he cast her as fickle and blamed her for not fulfilling her marital obligations. According to his version of events, Margarida never wanted to marry Johan. She had resisted and cursed those who arranged the union. She had called him names and threatened to run away. Only days before his death, Raymon alleged, Johan and Margarida had fought. Margarida fled to her family in Beaumont, bringing with her various possessions. Eight days later, she returned to Manosque. Instead of spending her first night back with Johan, however, she stayed in the cloister, and only returned home early the next morning. She told her brother that she had asked Johan for money to buy *menudetos*, honeyed treats; but Johan did not want sweets, and so they prepared a meal with garlic. When he finished eating, Johan left to work the fields. On his way, he complained to his servant that his tongue burned and asked who had prepared the meal. The servant did not know. As he sowed seeds, Raymon told the court, his brother began to experience pain in his stomach and chest. Johan's skin turned black, then red, and the servant told him to return home. When he arrived home, Johan went to bed and died shortly thereafter. Raymon's account of the events implies a relationship between the couple's fight and Johan's death that was more than coincidence. It was Margarida's behavior following her husband's death, however, that he found most suspicious.

According to Raymon, instead of tending her husband's corpse and mourning his passing, as would a decent wife, Margarida fled to the cloister. Guiltily, he said, Margarida doubted the cloister's protection and sought greater refuge in St. Anthony's Chapel. She did not participate in the rituals of death that commemorated her late husband; neither did she attend his burial nor weep. Equally significant, when the court messenger summoned her from the cloister to the court, she failed to appear. Surely, according to Raymon, these behaviors evinced her guilt.

Raymon's narrative of the events leading up to and following his brother's death was clearly meant to implicate her. Her alleged behavior failed to conform to the social norms of the dutiful wife. Unhappy in her marriage, it was assumed she poisoned or enchanted her husband's meal in order to escape.

Subsequent witnesses cast significant doubt on Raymon's allegations. These witnesses included a midwife, a neighbor, a servant, and a sister-in-law, and their words shed light on circumstances of Margarida's health that account for her suspicious behavior. Bila Fossata (the midwife) acknowledged that Margarida had, upon her return from Beaumont to Manosque, spent the night at the home of Antonieta Olivarie, a neighbor, but explained that she did so because of an illness. This illness formed the focus of Sanxia's testimony. Sanxia, another neighbor, told the court that Margarida claimed to suffer from an illness which the court scribe translated as *morbum caducum*, "the falling sickness"—the medieval term for epilepsy and chronic seizures. According to Sanxia, Margarida believed that this illness, which made it difficult for her to stand, began when she moved to Manosque. Margarida, herself, when questioned by the court, revealed that her illness was so debilitating that it prevented her from knowing her husband carnally.

The court also investigated the allegation of poisoning or sorcery. When asked if she had heard that Margarida had poisoned or enchanted her husband, Bila implicated Raymon as the source of such rumors. Raymon had, according to Bila, gone about yelling, "how did this happen?," and other things intended to inspire gossip. Some of the most exculpating evidence, however, came from the servant, Johan Baudiment of Volx. The servant recognized that Johan and Margarida sometimes fought, but added that they also joked. He also explained that Margarida had gone to Beaumont with her husband's blessing, and that, when she returned, it was Johan, her husband, who told her spend the night elsewhere. Most significantly, the servant described the allegedly enchanted or poisoned meal. He, Johan, and Margarida's brother, Bartomieu, had prepared the stew that day; Margarida had only added a bit of oil when she arrived and had done so in front of everyone. They all then ate the stew. The scribe noted that the servant then added, without being asked, that Johan had failed to finish his meal and had given the rest to his dog, who was still alive. As testimonies were usually shaped by the questions of inquisitorial procedure, the servant's voluntary addition of this event illustrates its significance to the accusations. The servant provided other illuminating details. Johan had been suffering, for some time, from chest pain, and often chewed ginger to alleviate it. Also, when Johan consumed the allegedly fateful stew, he had complained that the garlic burned his tongue. These details, though seemingly mundane, informed and shaped the expert testimony of Vivas Josep, the physician whom the court enlisted to investigate Johan's death.

In order for Raymon's allegations to inspire public rumor and attract the court's attention, they needed to adhere to common ideas and practices; they needed to ring true. The court's reception to an accusation which linked poison and sorcery, thus, reflects their conflation in the medieval imagination. Authorities found it difficult to distinguish between substances which were naturally poisonous and substances rendered poisonous by enchantment (Kieckhefer, 1976, 49); thus, as Jacques Chiffoleau has concluded, the boundaries between poison and enchantment were fluid ("les frontiers sont floues entre l'empoisonnment et l'envoûtement") (Chiffoleau, 1984). The Manosquin court blurred these borders on more than one occasion. In 1300, for example, they investigated the case of Matilde, a woman suspected of having used *maleficia sive fachuras...sive...causas venenosas* to kill her master and lover, with whom she had three children, after he married another woman (Shatzmiller, 1989). And, in 1321, public rumor brought Esmeniarde Bannilone and her daughter, Astrug, before the court on suspicion of having prepared *pocula seu comestiones venenosas...et plura alia maleficia in villa et valla Manuasce* (Courtemanche and Bednarski, 1998).

As the actors in these cases reveal, medieval culture typically considered poisoning and magic to be female practices, at least in the Midi, i.e., southern France. These were passive crimes, and lacked the honor of direct force by which most men were accused of having committed homicide in Manosque (Bednarski, 2014). Women, thought weaker by nature than men, were more vulnerable to the influence of evil forces that inspired sorcery than their male counterparts. Their knowledge of the secrets of nature and reproduction, furthermore, rendered them both more able and more likely to resort to magical or poisonous means. With the exception of the case against Margarida, most accusations of sorcery in Manosque reveal its employment for amorous or reproductive ends (Courtemanche and Bednarski, 1998), and Ghersi's research on poison and sorcery in the Midi confirms a blurring between women who poisoned and those who worked with herbs (Ghersi, 2009). When Raymon accused Margarida of using poisoning or sorcery to kill his half-brother, he operated within the gendered supernatural boundaries of his culture. As a woman suffering from epilepsy, however, Margarida was doubly vulnerable to these accusations. Temkin has illustrated that learned theories linked epilepsy to moral weakness in the premodern period, and the pathological state of the mental organs in epileptics allowed demonic forces to influence them more powerfully (Temkin, 1971).

Franck Collard, reflecting on the conflation of poison and sorcery in the medieval imagination, has argued that these two crimes were linked by their occult natures. Their indiscernible modes of action inspired a horror illustrated by the hyperbolic language used to describe them (Collard, 2003). Yet, as the testimony of Vivas Josep in the case of Margarida illustrates, medical authorities developed a system of identification and classification of poisons for theoretical and therapeutic reasons. These theories helped to recognize and counteract the effects of poison and to demystify its nature in medieval discourse.

The Manosquin court frequently summoned experts to testify in cases. As Courtemanche has described, the expert's purpose "was not to speak the truth—after all, such was the role of the witness. Instead, his statement was recorded in the register of credibility" (Courtemanche, 2002). From 1262 onward, physicians featured prominently amongst expert witnesses in Manosque (Shatzmiller, 1989). Thus, on 24 November, the court, at the behest of Margarida's brother, Peire de Portu, summoned Vivas Josep, a prominent Jewish physician in Manosque, to provide medical expertise on Johan's cause of death. Josep, along with the judge and notary, visited the deceased man's home to examine his corpse. First, he examined his lips, eyes, tongue, and the complexion of his face; then, he considered his nails, checking for discoloration, collected hair samples, scraped his scalp. He then turned to members of the household to reconstruct Johan's final moments. The servants explained that Johan had eaten a soup of garlic and almonds before leaving to work the fields, but had returned home because of a pain in his heart and died. Josep wanted to know if Johan had vomited. The servants confirmed that he had, but, when the physician asked to examine the vomit itself, they said that they had removed it.

Having concluded his examination, Josep turned to toxicological theory to explain the cause of death. He carefully situated his own conclusions within learned medical theories through the phrase "following the art of medicine" (*secundum artem medicine*) (Courtemanche, 2002), and cited specific authorities in the course of his testimony. He explained that there are many types of poisons, which Albucasis divides into three categories: animal, vegetable, and mineral. According to Avicenna, one can discern the nature of a poison by examining the deceased's vomit; although Josep lamented that he could not do this, he also noted that the Arabic physician had written that an individual who consumed poison will display certain signs, specifically, swollen lips or tongue, or protruding eyes. Josep observed neither of these signs on Johan's body, nor any blackness of the nails, and concluded that there were no signs of poison.

Josep provided further evidence that Johan could not have been poisoned. He recalled that Johan had eaten garlic for breakfast. According to what he called Lo Circainstans'[1] chapter on garlic, garlic inhibits poison. Josep further recalled Galen, in his *De Bone Digestione*, who states that milk, garlic, wine, vinegar, and salt dilute the effects of poison, and, his *De Regimine Sanitatis*, where he identifies garlic as a medicine that lessens and stops inflammation.

As the advice of learned authorities ruled out death by poisoning, Josep set about establishing a cause of death. He turned again to Galen and Avicenna for possible explanations of sudden death. He noted that, according to chapter 30 of Galen's *De Therapeuticorum*, syncope is dangerous and potentially lethal, since, as noted elsewhere in Galen, it destroys the spirit of life. Josep also proposed epilepsy as a possible cause, as it, too, can cause sudden death. Citing Avicenna, he noted that it came and went quickly, and that those who have it are some-times strong, sometimes weak. Josep ultimately decided, based on witness testimony, on syncope as cause of death. He concluded that the couple's inability to consummate their marriage due to Margarida's illness had caused a build-up of hot humors in her husband, and that this generated melancholy and ultimately led to syncope. Syncope, however, thought to have a cold complexion, reacted with the hot humors and suffocated the heart, causing it to tear. Complexion, in this sense, refers to Galen's description of the qualities hot, wet, cold, and dry. Indeed, to support his conclusions, Josep, again, cited Galen and Avicenna.

Vivas Josep's medical testimony reveals the application of a learned theory of toxicology that sought to identify and understand poison in the Middle Ages. Poison challenged authorities because its effects could imitate genuine illnesses (Collard, 2008), as the case against Margarida reveals. Galen attempted to incorporate earlier, individual

---

[1]Courtemanche has identified this as Mathieu Platearius' *Circa instans* or *Liber de simplici medicina* (Courtemanche, 2002).

descriptions of poisons into a theoretical framework by integrating toxicology into his pharmacological theory. Thus, he understood poisons by their complexion. Their effects, however, seemed to differ little from those brought on by medications or illness. Avicenna subsequently elaborated on Galen's theories through the introduction of the concept of "specific form," a mode of action alongside complexion. Chandelier reports that, according to Avicenna, specific form was a form acquired by a substance after complexion; it derived from the mixture of these simple elements which characterized the complexion. From this mixture came an aptitude or predisposition to receive a species, a form added to that which the simple elements already possessed. It was a sort of perfection acquired by the matter in consequence of this aptitude derived from complexion. ("C'est une forme qu'acquiert une substance postérieurement à la complexion; elle dérive du mélange des éléments simples qui la caractérisent. De ce mélange résulte une aptitude à recevoir une espèce, une forme ajoutée à celle que les simples possèdent ... C'est une sorte de perfection acquise par la matière en fonction de cette aptitude dérivée de la complexion.") In the 14th century, Cristoforo degli Onesti, following on Avicenna, rejected complexion theory in favor of specific form as the sole mode of action. There then appeared a theory of toxicology that clearly distinguished poisons from drugs (Chandelier, 2009). Vivas Josep's methods and theoretical applications thus reflect the learned attempts of the time to understand poisons.

The accusations and investigations contained within the trial of Margarida de Portu reflect the place of poison in the medieval imagination. That the ambiguous suspicion of poison or sorcery that Raymon Gauterii planted stirred the court to hear the case reflects the conflation of these two crimes in medieval society. They were bound together by gender assumptions that linked poison to a feminine, occult act. But Vivas Josep's investigation and testimony underscore how in Provence, by 1394, learned medical discourses were beginning to reject an occult perspective to toxins and instead approached poisons and poisoning from an evidentiary perspective, grounded in anecdotal examination and supported by the written expertise of physicians.

## REFERENCES

Bednarski S: *A poisoned past: the life and times of Margarida de Portu, a fourteenth-century accused poisoner*, Toronto, 2014, University of Toronto Press.

Chandelier J: Théorie et définition des poisons à la fin du Moyen Age, *J Med Hum Stud* 17:23–38, 2009.

Chiffoleau, J., Sur la pratique et la conjuncture de l'aveu judiciaire en France du XIIIe siècle au XVe siècle, in *L'Aveu, Antiquité et Moyen Age, Actes de la Table Ronde organisée par l'Ecole française de Rome (mars 1984)*, 1986, École Française de Rome; Rome, 341-380.

Collard F: *Veneficius vel maleficius*. Réflexions sur les relations entre le crime de poison et la sorcellerie dans l'Occident médiéval, *Le Moyen Age* 111:9–57, 2003.

Collard F: *The crime of poison in the Middle Ages*, Deborah Nelson-Campbell, London, 2008, Praeger Publishers.

Courtemanche A: The judge, the doctor, and the poisoner: medical expertise in Manosquin judicial rituals at the end of the fourteenth century. In Koster JR, editor: *Medieval and early ritual: formalized behaviour in Europe, China, and Japan*, Leiden, 2002, Brill, pp 105–123.

Courtemanche A, Bednarski S: De l'eau, du grain et une figurine à la forme humaine. Quelques procès pour sortilèges à Manosque au début du XIVᵉ siècle, *Memini. Travaux et documents publiés par la Société des études médiévales du Québec* 2:75–105, 1998.

Ghersi N: Poisons, sorcières et lande de bouc. In Collard F, editor: *Le poison et ses usages au Moyen Age*, 17, 2009, Cahiers de recherches mediévales, pp 103–120.

Kieckhefer R: *The European witch trials: their foundations in popular and learned culture, 1300–1500*, Berkley, 1976, University of California Press.

Shatzmiller J: *Médecine et justice en Provence médiévale. Documents de Manosque, 1263–1348*, Aix-en-Provence, 1989, Publications de l'Université de Provence.

Temkin O: *The falling sickness: history of epilepsy from the Greeks to the beginnings of modern neurology*, Baltimore, 1971, Johns Hopkins University Press.

# CHAPTER 15

# Animal Venoms in the Middle Ages

**Kathleen Walker-Meikle**
University College London, London, United Kingdom

## 15.1 INTRODUCTION

Accounts of bites by venomous animals appear in numerous medieval historical and literary sources, although this chapter will focus on learned medical and natural history texts. Bites by venomous animals, notably snakes, scorpions, spiders, and rabid-dogs, were considered to be very distinct from wounds inflicted by other means and were believed to cause severe physical and mental symptoms for the patient, including derangement, hallucinations, or bloodcoming out of every pore. In medical literature on animal bites that discuss symptoms and therapeutics, venomous animals receive the lion's share of attention, and if the animal could not be identified as venomous or not, most authors suggested erring on the side of caution and treating the bite as venomous. Physicians and surgeons in various treatises suggested the use of a wide variety of medical treatments. Unguents, plasters, syrups, assorted theriacs, purgatives, and special diets were all prescribed, along with the use of ligatures, scarification, opening of wounds, caustics, cupping, cauterizers, and sucking the wound. Sea-bathing was praised as both as a cure and as a prophylactic against rabies, while fumigations and other methods to ward off small biting animals were also suggested. Animal bites of all kinds were believed to contain noxious poisons that needed swift attention, and thus were clearly distinguished from wounds caused by other means. Serpents and rabid dogs were the two most feared venomous animals. The saliva of a rabid dog was considered to be a poison and to be as dangerous as that of the most deadly snake. Animal venom was considered to be a type of poison, along with poisons of mineral origin, botanical origin, and animal origin. Poisons of animal origin were substances that might be consumed (e.g., eating the brain of a cat could provoke madness) while venom involved the bite or sting of an animal (d'Abano, 1473).

Toxicology in the Middle Ages and Renaissance. DOI: http://dx.doi.org/10.1016/B978-0-12-809554-6.00015-9

## 15.2 VENOMOUS SNAKES

Venomous snakes occupy a major part of the medical literature on poisons. The hugely influential Canon of Medicine by Avicenna (Ibn Sīnā, 980–1037) combines Classical, Byzantine, and Arabic sources in the section on animal bites, a subcategory of poisons in the text (Book 4 Fen 6, 98 chapters). Following natural history conventions of the time, the main division is between "crawling animals" and quadrupeds. The former covers serpents of all kinds (15 in total), scorpions, fleas, wasps, bees, flying ants, spiders, lizards (including "the animal with forty-four feet"), the "Red Sea frog," newts, salamanders, and assorted venomous marine reptiles. The category of quadrupeds includes wolves, weasels, mice, apes, cats, crocodiles, and both rabid and nonrabid men and dogs.

Avicenna's text was adopted and adapted by Latin medical scholars despite the different environmental realities and species coverage. Some authors would focus their discussions on snakes that abound in their geographical area. Thus, Bernard de Gordon, Professor of Medicine at the University of Montpellier from 1285, would remark that while the tyrus, dragon, asp, and basilisk were deadly, particularly the basilisk as it could kill with its sight and even affect birds flying above, they did not reside around Montpellier, and that most of the local snakes were not highly venomous (de Gordon, 1497).

Similarly, the early 14th century French surgeon Henri de Mondeville remarked that the most venomous of serpents were never seen in France, local spiders were quite harmless, a few nonvenomous lizards might bite, scorpions lived in Italy and Avignon, and that the major cause of snake bite was due to people dragging grass snakes out of their burrows and shoving them in sacks or harassing them. For Mondeville, these grass snakes (*Natrix natrix*) were common in houses but would only bite if injured or challenged. Even if they bit someone, the wound would be superficial and easy to treat (Henri de Mondeville, 1893).

### 15.2.1 Albertus Magnus and Venomous Serpents (13th Century)

An encyclopedic coverage on venomous snakes appears in the Dominican scholar Albertus Magnus's (d. 1290) *On animals*, a long commentary on Aristotle's natural history (d. 1290) (Albertus Magnus, 1999). His major source on snakes was Avicenna, but other sources included Pliny the Elder's Natural *History*, Thomas de Cantimpré's *On*

*Natural Things*, Isidore of Seville's *Etymologies*, and Jacques de Vitry's *History of the Orient*. Regarding the nature of snake venom, Albertus (following Avicenna, Canon 4.6.1.3), divided snakes into three groups according to the "operations of their poisons." The first venom of the first group had "violent sharpness," and there was usually no cure, apart from immediate amputation or deep cauterization of the affected limb, with death following in 3 years. Serpents in this class included the *regulus* or *basilicus*, the *yrundo* (with the coloring of a swallow) the dry asp, and the *exspuens* (the spitter). Their bites were considered "mute" as they would hardly be noticed before causing damage. These serpents bend themselves into the shape of a bow before striking, and many of the Egyptians species had two horns (a likely reference to *Cerastes cerastes*).

The second group's venom was of "moderate sharpness" and usually killed in 1 − 7 hours. Vipers made up most of this group. The third group involved a venom of little strength that did not require specific treatment, although the bite itself might produce ulcers and swellings. In addition, Albertus remarked that the "sharpness" of the snake venom was affected by various factors. These included the sex of the snake, the age (old ones being more venomous), the environment (desert snakes being more dangerous than those living on riverbanks), the passions (the venom of angry snakes being worse than that of young ones), whether the snake had eaten recently (as hungry snakes produced a more toxic venom), and finally the weather, as the same species is more dangerous in summer than in winter. Heat always made all kinds of venom more potent, which is why rabid dogs, with their "venomous saliva," were most dangerous in the "dog days of summer" (connected to the rising of the dog-star Sirius). Snake venom was considered in humoral terms as a "hot" poison, in contrast to scorpion venom, which was "cold."

Albertus listed 61 different types of snakes, including the *horned asp* whose bite provoked loss of reason and a pain that resembled nails being driven into the site of the bite, the *cerystalys*, with symptoms including bloody black fluid and loss of reason before the fatal spasm, the *dypsa* which caused the victim to drink so much water that it would flow through their veins, the *basilisk*, which killed with its gaze or liquefied with its bite, and the *asfodius*, which caused blood to drip from all pores (an interesting interpretation of hemorrhagic venom), with the victim having convulsions before death.

The last snake discussed was the viper (*Vipera berus*), which would have been the most common venomous snake in Europe. Taking his cue from Avicenna (Canon 4.6.3.32), Albertus remarked that the bite of a viper was caused by two or three teeth, and that the site would first issue blood and then a hot, slimy fluid. The blood would be at first watery, then frothy, and then the color of copper rust. A warm red abscess would appear on the site of the bite, and later turn green. The patient would experience dizziness, lose their color, and usually die within 3–7 days.

## 15.2.2 Venom and Henri de Mondeville (14th Century)

Avicenna's influential text was supplemented by the *Treatise on Poisons* by Moses Maimonides (Moses ben Maimon, d. 1204), and translated three times into Latin by the late 13th and early 14th centuries. Following Maimonides, the French surgeon Henri de Mondeville (d. 1316) remarked that the general sign of a venomous bite was a half-dead patient who appeared comatose as if drunk or asleep, and had a painful burning wound. He stressed the importance of identifying the animal responsible, in order for the medical practitioners to tailor a suitable treatment. The best case scenario would be if the patient could identify what animal had bitten him. If this was not possible, then the attending physician or surgeon would have to examine the wound for clues. For example, a viper's bite could be recognized by the green skin at the site of the bite, and it was possible to even identify if it was a male or female viper, as there would be more than two punctures in the skin in the latter case, and this was important as the venom of the male viper was considered to be worse (Mondeville claimed that they bit with their side teeth, not those at the front). Similarly, a scorpion's puncture could be identified by stone-hard edges accompanied by a slight swelling, while a bee's stinger would be left in the puncture site. In general, patients bitten by venomous animals would be comatose and the wound would have a painful burning sensation, in comparison to wounds made by other sources.

Apart from therapeutics, Mondeville also offered preventative measures, such as avoiding, if possible, venomous animals, particularly if they were angry or if it was hot. Similarly, he recommended not injuring or exciting them or standing in their escape path. If someone needed to travel to a region full of venomous snakes, they should take a theriac of nuts as a prophylactic. Fumigants could also be used, and the smoke of a burning deer's antler was particularly recommended against venomous snakes.

### 15.2.3 An Early 14th Century Case of Snake-bite (Gentile da Foligno)

The *Consilium regarding the bite of a deaf asp* details an actual case of venomous snake bite and treatment in the first half of the 14th century. The author and physician was Gentile da Foligno (d. 1348) who was trained at the universities of Padua and Bologna and at the time of the case was teaching at the University of Perugia (Thorndike, 1961).

A young man on a mountain near Perugia had been bitten on the ankle by a snake. The young man had made a tourniquet himself but by the time Foligno was called a few hours later, he was motionless with closed eyelids. Foligno, doubtful of the success of his efforts, began applying the Great Theriac (a compound medicine against snake bites and other ailments) topically onto the chest in the region of the heart, gave him stale butter to drink, and started scarifying and cupping the wound. The snake, now dead, was bought to the doctor to be identified, who described it as a "short deaf asp, little more than a cubit in length, black with gray spots, with a broad head, and a short tail" (likely an asp viper, *Vipera aspis*). After not being able to find volunteers to suck the venom out with their mouths due to the patient's "horrible" condition, Foligno gave the patient a mixture of gentian, balsam seed, rue, and anise in wine to drink. The next day, the patient was still in bed, with a high pulse and a face "foul in color and form." Foligno gave the patient emerald power and wine infused with citrus seeds (both recommended by Maimonides) although the patient could hardly swallow. That afternoon, the patient was still in poor condition, and his urine was inspected and found to be red. In consultation with other doctors, Foligno decided to try a theriac medicine recommended by Haly Abbas ("Ali ibn al-" Abbas al-Majusi, d. c. 982−994) which contained castoreum (substance from the castor sacs of beavers), cassia wood, citrus seeds, anise seed, pepper, and aristologia. On the evening of the second day, the patient could recognize his doctors and praise God, despite still having a high pulse and red urine. Foligno prescribed more emerald powder, Haly Abbas theriac, and citrus seeds, along with massages.

In the early morning of the third day, the patient appeared to have taken a turn for the worse and could not see, and passed a thin livid urine full of "scaly solutions." Foligno heard that the patient had taken hardly any of the Haly Abbas theriac and had slept deeply. Emerald powder, citrus seeds, and more Haly Abbas theriac were now forcibly administered, and when visiting the patient later, Foligno found him much improved, and ordered a diet of chicken fat and chicken soup (with emerald powder and citrus seeds sprinkled on top). That evening, more theriac was prescribed, cupping glasses were put on the wound and orders given that the patient could not be allowed to sleep too much. By the fourth day, Foligno found that the patient's urine was of a better color although the young man complained of stomach and kidney pains. An enema of milk mixed with a decoction of mallow was administered, along with body wash with an infusion of *Aristologia rotunda*.

By the fifth day, the patient had no grave symptoms whatsoever, a new diet of good food and wine was prescribed, along with the suggestion of eating sour things and taking "hot" theriacs to counter the "hot" snake venom, as the patient appeared to be slightly deranged, likely a symptom of the venom. All in all, a success, as Foligno remarked that people bitten by this species of snake often do not open their eyelids for months, ending his treatise on a positive note: "So that physicians consider the marvelous effects of poisons and the marvelous properties of things against them, and in all these see the beauty of glorious God."

## 15.3 SCORPIONS, SPIDERS, AND OTHER "VENOMOUS" ANIMALS

In Albertus' *On Animals*, the chapter on serpents also included the deadly *rutela* spider, whose bite involved great pain, with the patient turning yellow. A substance similar to a spider web was seen to come out from the site of the bite, and the patient might on occasion urinate with great pain small pieces of flesh in the shapes of spiders.

In the same work, the "cold" venom of the scorpion was in marked contrast to the "hot" venom of serpents. Albertus described the pain to being akin to being crushed by "salt-crushing pestles," and the victim would vomit matter that would instantly congeal. Viscous fluid would come from their lips and eyes, and there would be a notable change to their complexion. Henri de Mondeville (early 14th century) recommended using general medicines and topicals against venoms on scorpion stings, although he noted that garlic and asafetida in wine was good to drink and that an easy remedy involved boiling some scorpions in oil and using it to anoint the puncture site. Regarding spiders, Mondeville noted that there were no venomous spiders in northern France and that the common spider bite was hardly venomous at all. On lizard bites, he remarked that they were mostly nonvenomous, and the only concern was that their teeth might be left in the wound, for which he suggested rubbing the wound with oil until the teeth came out. Bernard de Gordon notes that scorpions were common in Avignon and other regions, and that their "cold" poison could be treated with theriac taken either internally or placed topically on the puncture. Bees and wasps, in contrast, had a "hot" burning sting, and he recommended placing a cold piece of iron or lead chilled in vinegar on the puncture site.

The physician William of Marra wrote an extensive treatise on poisons for Pope Urban V (titled *The Papal Garland*, 1362) and along with the familiar snakes, scorpions, and other venomous beasts, included a chapter on the tarantula. Tarantula bite could be treated by playing music to the patient, as its venom caused melancholy (an excess of black bile), and music could produce joy and prevent the venom from penetrating the patient's vital organs. William of Marra dismissed the explanation of the "vulgar" who claimed that it was due to the tarantula itself singing as it bit the victim, who himself could be treated by hearing similar music. This narrative is likely one of the earliest sources on Southern Italian tarantism, a form of hysteria resulting in uncontrollable dancing and believed to be caused by the bite of a tarantula (Thorndike, 1934).

Bites and punctures, from both venomous and nonvenomous animals, appeared frequently in medical and laid sources throughout the High and Late Middle Ages, understandably, in a society where humans lived in close contact with many animals, both wild and domestic. This chapter has offered a brief overview of the medical category of bites and punctures as described in learned medical literature and has elaborated on some of the main theoretical and practical ideas regarding animal toxicity, and how scholars and medical practitioners understood the causes, symptoms, and treatment of animal bites.

## REFERENCES

Pietro d'Abano: *De venenis* (Padua, 1473).

Bernard de Gordon: *Lilium medicinae* (Venice, 1497). Book I, chapters 14–17.

Henri de Mondeville: *Chirurgie de maitre Henri de Mondeville, chirurgien de Philippe le Bel, roi de France, composée de 1306 à 1320, trans.* E. Nicaise (F. Alcan: Paris, 1893), pp. 436−457. Also see Maimonides, Moses: *On poisons and the protection against lethal drugs,* ed. and trans. G. Bos and M. R. McVaugh (Brigham Young University Press, 2009).

Albertus Magnus: *On Animals: A Medieval Summa Zoologica, trans, vol. 2.* In Kitchell KF, and Michael I, editors, Baltimore, 1999, The Johns Hopkins University Press, pp. 1709−1738.

Thorndike, L: *A history of magic and experimental science* (vol III), New York, 1934, Columbia University Press, p. 534.

Thorndike L: A case of snake-bite from the consilia of Gentile da Foligno, *Med Hist* 5(1):90−95, 1961.

# CHAPTER 16

# Medical Literature on Poison, c. 1300—1600

**Frederick Gibbs**
University of New Mexico, Albuquerque, United States

Medieval and Renaissance toxicology, while of course indebted to classical frameworks, took up distinctly new ways of inquiring about the nature and properties of poisons and venoms. Although historical conceptions of poison can and should be approached from numerous perspectives, this chapter outlines the broad contours of late medieval and early modern toxicology as seen from some widely cited Latin medical texts on poison composed between 1300 and 1600 (Frederick W. Gibbs, 2017).

Before addressing the Latin texts, it is necessary to highlight the *Canon* of the great Persian philosopher and physician Avicenna (fl. 11th cen.), which became foundational for toxicology in the Latin West. Avicenna stated that whatever substances change the body and are changed by the body (but not assimilated to it), are poisonous medicines; whatever substance is not changed by the body but instead changes the body to itself (i.e., converts the body into that substance) is an absolute poison (Avicenna, 1564a). He also clearly identified two species of poison, differentiated by their operative power: One species of poison operated by virtue of its complexion or qualities (its makeup of hot, cold, wet, and dry), which is largely how classical Greek physicians like Galen conceived of poison; the other species operated by virtue of its specific form (*forma specifica*), or total substance (*tota substantia*), a conception that Avicenna emphasized far more than his predecessors (Avicenna, 1564b).

Avicenna's description of poison was gradually incorporated into later medieval encyclopedias (both medical and general ones), which usually included a section on venomous animals, but did not focus on *venenum* itself (both here and throughout this Chapter 1, Poison and Its Dose: Paracelsus on Toxicology use *venenum* to signify not merely venom, but poison in general, as did Latin physicians). A Latin compilation of such poison lore appeared in the late 13th century by the Castilian Franciscan

Toxicology in the Middle Ages and Renaissance. DOI: http://dx.doi.org/10.1016/B978-0-12-809554-6.00016-0

friar Juan Gil of Zamora (Johannis Aegidius Zamorensis, 1241 – 1328). His *Liber contra venena et animalia venenosa* (*Book against Poisons and Venomous Animals*) was composed between 1289 and 1295 and, following classical precedent, presented 19 alphabetically ordered chapters listing animal, plant, and mineral antidotes for various poisons (usually from venomous animals) (de Zamora, 2009). Juan's *Liber contra venena* emerged from the encyclopedic tradition and its principal sources, namely Pliny, Isidore, Gilbertus Anglicus, and most heavily from Vincent of Beauvais. While focused on listing antidotes, Juan also addressed topics such as precautions for avoiding poisons and how to recognize general signs of poisoning; as did most other encyclopedists, he also echoed Avicenna's statement that poison operates either by its qualities or by its total substance. Although Juan's text does not seem to have been known (or of interest) to later authors of poison texts, it certainly testifies to the interest in consolidating the growing body of knowledge of poison in the late 13th century.

The nature of *venenum* itself received considerably more attention from Pietro d'Abano (d. c. 1316), well known as "The *Conciliator*" for his monumental work of the same name that tried to reconcile some points of apparent conflict between Aristotelian philosophy and Galenic medicine. Perhaps even more widely circulated than his *Conciliator* was his treatise *De venenis et eorum remediis* (*On Poisons and Their Remedies*). Following classical precedent, Pietro provided a list of dangerous substances, their signs, and their antidotes. Although not entirely novel in its formulations, what makes Pietro's *De venenis* so crucial in the history of toxicology is how Pietro undertook a sustained and distinctive philosophical inquiry into the nature of poison in a standalone medical text. Following Avicenna, Pietro immediately established his position on the nature of poison (which would become one of his most cited passages), stating that "poison is the opposite of food with respect to our bodies" (d'Abano, 1924). In other words, food nourishes, poison destroys. It was likely a combination of Pietro's clear elucidation and elaboration of earlier authors (particularly Avicenna), the way he drew attention to *venenum* in a separate text (i.e., not buried in a large medical encyclopedia), and his own reputation that helped his *De venenis* become so influential. Judging from later references to him and his work, Pietro quickly became a preeminent authority on poison in the Latin West.

Just over three decades after Pietro's treatise, the so-called Black Death provided yet another inflection point in the history of toxicology.

Although not writing specifically about poison, some physicians argued explicitly that *venenum* was a cause of pestilential disease. Prior to the outbreaks of plague in the mid-14th century, conceptions of poison as a cause of disease centered on envenomations and poisonings from toxic plants. However, even in these cases, physicians generally wrote about the overwhelming heat and cold of these poisons, emphasizing Galenic complexional imbalance as the cause of disease. In contrast, plague tracts show that the cause of pestilential disease was at times clearly and distinctly identified with *venenum*—and its power derived from its specific form—engendering a new and significant role for poison in disease causation and dissemination.

One novel and lasting characteristic feature of plague treatises was the extent to which physicians described the air as either being poisonous or containing a poison—and thus acted as a causal agent in spreading the disease. A typical formulation comes from Jean de Tournemire (1329–96), a papal physician who studied at Montpellier and who described pestilence as "a poisonous and infected air from corrupted vapors and poisonous exhalations" (Jean de Tournemire, 1911). More significantly, this poison found its way into the body and was directly responsible for symptoms of plague. Gentile da Foligno, a student of Pietro d'Abano, wrote in his plague tract from the late 1340s that when the corruption (from the air) entered a body, a "poisonous matter" (*materia venenosa*) was generated near the heart and lungs and acted as the most immediate cause of disease. This material did not act by means of its qualities (too much heat or cold, for instance) but through a property of being poisonous (*per proprietatem venenositatis*) (Gentile da Foligno, 1515).

Moving to the later 14th century, several northern Italian physicians such as Cristoforo de Honestis and Guglielmo de Marra eagerly took up new inquiries into the nature of poison. Guglielmo, for instance, extended Pietro's natural philosophical inquiry into the causes and effects of poison with a set of 14 questions concerning how and why poison behaved in certain ways, particularly the potential processes of poisoning. Many of these concerned the effects of poisons from certain animals, such as whether a basilisk (a mythical serpent well known to kill through all five senses) could kill a man by sight, whether a tarantula can kill by sound (no; it merely causes melancholy), and why a bite from a rabid dog near the face is usually fatal (because it is nearest the sense organs) (Biblioteca

Apostolica Vaticana). These physicians were asking new questions about poison, as well as asking new *kinds* of toxicological questions that increasingly shaped the conception of poison itself.

Fifteenth-century toxicology is anchored by the works of two Italian physicians, Sante Arduino and Antonio Guaineri, who provided two complementary approaches to understanding poison. In contrast to earlier authors of poison texts, Guaineri (who completed his *De venenis* around 1440) was more interested in outlining the properties of poison and poisonous bodies than establishing a specific definition of *venenum*. What was the nature of a poisonous body? Did the condition of being poisoned pose a threat for other bodies nearby? Could a property of being poisonous be transferred across space, or by contact alone? Guaineri's natural philosophical inquiries led him, for example, to wonder if a deaf and blind man would be immune to the basilisk's unquestionably poisonous nature. To answer, Guaineri questioned the notion that a basilisk must poison exclusively by either sight or poisonous vapor, as well as whether a poison could propagate itself effectively over distance. Regardless, Guaineri concluded, "... someone seen by a basilisk is not infected to the point of dying" (Guaineri, 1517). There were, of course, neither easy nor agreed upon answers to such questions. But the answers are much less important than the nature of these discussions, particularly how they illustrate the continued development of a new toxicology that began in the 14th century, including a focus on the nature of *venenum*, as well as using the model of *venenum* to understand the nature of disease and contagion.

In the 1420s, Sante Arduino completed an eight-book compendium on poison, undoubtedly the most textually ambitious of any poison treatise produced between the 14th and 16th centuries (Arduino, 1562). Totaling nearly 500 printed pages, this massive encyclopedia canvassed a seemingly interminable list of poisonous plants and minerals, as well as various categories of venomous bites (such as those from violent animals, reptiles, fish, and four-legged animals). Far surpassing what earlier writers of poison tracts had summarized, Arduino cited over 40 different (mostly medical) authorities, spanning a wide range of medical literature.

Arduino's text uniquely combined a natural (and textual) history with the natural philosophy of poison, foreshadowing the relationship between natural history and medicine that would blossom in the 16th century. Similarly, it also illustrates the transition between the medieval interest in true and universal knowledge acquired through axioms and syllogistic reasoning and later interest in and descriptions of particulars. Arduino bridged theory and practice by employing universals to understand how poison as a category of substance worked inside the body (the *scientia* [knowledge] of poison); at the same time, he elaborately documented the particulars of many dangerous substances—borrowing from dozens of earlier writers—in order to enable recognition and treatment of individual cases of poisoning (the *ars* [art] of treating poison victims).

Over the course of the 16th century, physicians fundamentally reframed their debates about the nature of poison and formulated new ways of discussing it. One motivation stemmed from a tension between the Greek *pharmakon* and the Latin *venenum*, a discrepancy that emerged from Renaissance efforts to retranslate Greek texts into Latin. In the Greco-Latin version of Dioscorides's *De materia medica* (c. 1518), for instance, the Florentine humanist Marcello Virgilio Adriani (1464–1521) lamented the difficulty of translating *pharmakon*, "which the Greeks used indistinctly for speaking of both helpful medicines and of lethal poisons, but for which the Latin language has not provided the equivalent to suggest the double meaning" (Adriani, 1518). Partly because of the linguistic and conceptual tension between *pharmakon* and *venenum*, and partly because of the expanding natural world, physicians endeavored to understand the concept of poison in the larger order of nature and an ever-increasing number of medicinal substances.

As a result, 16th-century physicians approached *venenum* in two ways that were distinctly different from their predecessors. First, they emphasized that it was impossible to understand poison as either a universal or a particular phenomenon without accounting for both perspectives simultaneously. Physicians consequently crafted a new concept of *venenum* defined dually: One axiomatically by its universal ability to harm the human body (and thus curable through universal remedies); the other by emphasizing the particular classes, properties, and natures of various poisons. Second, they wrote more explicitly and extensively about the relationship between poison and disease.

The famous Paduan physician Girolamo Cardano (1501–76), for instance, in his text on poison from 1564, adopted a universal definition of *venenum* that deliberately encompassed many specific kinds of poisons, ways they affected the body, and processes of poisoning. Cardano stressed the physical qualities of poison that made it poisonous rather than simply harmful, similar to but different from the Avicennean *forma specifica* that previous authors typically employed in their works on poison. It was some kind of delicate substance—not merely an occult quality—that could turn a harmful medicine into a poison. Cardano expressed the necessity for physicians to understand the individual natures of a wide spectrum of poisons, not just for theoretical, but for practical applications, because "although it is with the greatest difficulty, it is necessary for a physician to know poisons and their circumstances, including their manner [of action], strength, species, and quality, if he wants to treat correctly" (Cardano, 1653b).

Another approach to refining the notion of poison comes via Girolamo Mercuriale (1530–1606), who composed his treatise *De venenis et morbis venenosis* (*On Poisonous Diseases and Poisons*) in 1584. Mercuriale focused on reconciling ancient and contemporary definitions of poison that straddled the universal and particular. He carefully emphasized, for instance, the spectrum of potential harm of the great variety of so-called poisons: "although all poison by its nature harms and corrupts the nature of man, the difference [between particular poisons] is not small, since some never harm, even if they have the potential to do so, and other always do, but some slowly, others quickly"(Mercuriale, 1584a). He went on to explain how some so-called poisons may not harm at all, depending on the body's individual properties or if the body has been prepared ahead of time to withstand the poison, such as in the legendary tale of Mithridates, or if the poison is removed from the body quickly enough. His comparative approach helped him to outline the difference between venom and other poisonous substances, stating that the venom of venomous animals has a natural sympathy to itself, and thus can be used to extract poison from the body, whereas other poisons are more liable to injure the body.

Unlike earlier literature, the relationship between poison and disease featured prominently in 16th-century texts on poison. Cardano focused on the relationship between poison and putrefaction, noting that the kind and severity of disease was determined by the level of perfection that the putrefactive process obtained (Cardano, 1653a). It was in fact according to various corruptive processes rather than the traditional tripartite division of animal, vegetable, and mineral poisons that guided Cardano's chapters on the many possible origins of poison, whether from putrefaction or corruption (both inside and outside the body), and whether from air, water, contagion, or by another poison. In this way, similar to earlier plague authors, Cardano implicitly alludes to what we now consider issues related to environmental health and pollution, namely how toxins or corruptions in the environment can invoke a kind of corruption inside the body that directly leads to disease. Cardano's several long chapters on the nature of putrefaction and corruption certainly stand out as unique features for a text on poison. While Cardano did not say much about poison itself in these sections, it is clear that he wanted to establish the centrality of the role of putrefaction in generating poison inside the body and the kinds of disease that may arise from it. Mercuriale, in contrast, endeavored to restrict the notion of diseases caused by *venenum*, arguing across his first three chapters that not all diseases should be called poisonous. That which is properly called a poisonous disease, he argued, "is disposed against the nature of the heart, immediately attacking it, with swift action by virtue of its poison-like force" (Mercuriale, 1584b).

Although there was no formalized discipline of toxicology in the Medieval and Renaissance periods, and many ways that physicians conceptualized *venenum* were eventually replaced or abandoned, one can hardly deny the persistence of critical thinking about the nature of poison and its relationship to other substances, disease, and the human body. Whether or not any particular development should be said to constitute a definitive step toward more modern scientific concepts is far less important than appreciating the long and complex history of how physicians debated, from various perspectives, the nature of poison and developed a genre of medical writing that remains instrumental to the history of toxicology.

## REFERENCES

Adriani, commentary on Dioscorides, *De materia medica*, VI.0 (Florence, 1518, f. 332r): Anceps in graecis vox illa *pharmakon*: qua ea gens ad venena et morborum medicamenta utitur ... non respondente illi in latinis ancipiti aliqua voce alteram eius significationem ab altera discrevisse oportebat.

Arduino, *De venenis* (Basel, 1562).

Avicenna, *Canon*, IV.6.1.2 (ed. Casteo and Mongio 1564b, 191.B64—192.A6): Species venenorum sunt duae. Est faciens operationem suam cum qualitate, quae est in ipso: et efficiens cum forma sua, et tota substantia sua.

Avicenna, *Canon*, I.2.2.15 (ed. Casteo and Mongio 1564a, 104.B18—20): Illud vero quod a corpore nullo modo mutatur, et ipsum mutat; est venenum absolute.

Biblioteca Apostolica Vaticana, *Barb. lat.* 306, pp. 1–157.

Cardano, *De venenis*, I.7–8 (Padua, 1653a, 13–16).

Cardano, *De venenis*, I.2 (Padua, 1653b, 6): Difficultas maxima est in hoc, quod necesse est medico cognoscere venena & quae circa ipsa sunt, modos, vim, speciem, qualitatem, si debet recte curare.

Frederick W. Gibbs, More detail and analysis will soon be available in my forthcoming book, *Poison, Medicine, and Disease in Late Medieval and Early Modern Europe* (Routledge, Fall 2017).

Pietro d'Abano, *De venenis*, I (Marburg, 1537, 2; tr. Brown 1924, 27): Venenum oppositum est cibo nostri corporis.

Juan Gil de Zamora, *Liber contra venenia et animalia venenosa* (ed. Ferrero Hernández 2009).

Gentile da Foligno, *Consilium contra pestilentiam*, 2 (Salamanca, 1515, sig. a3r): Et causa immediata et particularis est quadam materia venenosa, quae est circa cor et pulmonem et ibi generatur: cuius impressio non est per excessum qualitatum primorum in gradu, sed propter proprietatem venenositatem.

Guaineri, *De venenis*, I (Lyon, 1517, f. ccxxviii[r].B5–9): Et isto modo ex quo vapor basilisci in isto casu hominem ipsum videntem at tingere non potest homo primo videns basiliscum ipsum perimit: qui in postea a basilisco visus non inficietur totaliter ut moriatur.

Jean de Tournemire, *De epydemia* (ed. Sudhoff 1911, 49): ... videtur mihi aer infectus et venenatus per vapores corruptos et exaltationes venenosas.

Mercuriale, *De venenis*, I.5 (Venice, 1584a, f. 7r): Et quamquam omne venenum suapte [sic] natura evertat, et corrumpat humanam naturam, differentiae tamen est non exigua, quia aliqua quidem nunquam interimunt, etiam si habeant potentiam interimendi: aliqua vero interimunt, sed aut tardius, aut citius.

Mercuriale, *De venenis*, I.1 (Venice, 1584b, f. 1v): Definitur autem morbus perniciousus hunc in modum: morbus venenosus est dispositio praeter naturam cordis, offendens immediate, et vehementer actiones ob vim veneficam.

# INDEX

Crawling animals, 152
Criminal magical underworld, poisons and,
    65–67
Cuneiform tablets, 119–120
Curative powers, 102
*Curcuma zedoaria.. See* Zedoary
    (*Curcuma zedoaria*)

**D**
da Foligno, Gentile, 155–156
Damascenus, Janus, 120
*De Bone Digestione*, 147
de Gordon, Bernard, 152
de Honestis, Cristoforo, 161–162
de La Reynie, Nicolas, 135–136
de Marra, Guglielmo, 161–162
*De materia medica*, 47–48, 50
*De methodo medendi*, 47
de Montespan, Marquise, 135–136, 139–140
*De Re Metallica*, 85–86
*De Regimine Sanitatis*, 147
*De Therapeuticorum*, 147
de Tournemire, Jean, 161
*De venenis*, 43–44, 50–51, 160
    Pietro d'Abano, 43, 45–46, 46*f*
Deceptiveness, 55–56
"Della Pietra Cerulea", 77
Della Rovere, Cardinal Giuliano, 54
Deshayes, Catherine, 136–137
Di Adamo, Teofania, 63–64
Dioscorides, 45–46, 50–51
    Pseudo-Dioscorides, 48
    *On venom* and *On poisons* to, 50–51
Doria, Prince Andrea, 71
Draconites, 119–120
Drugs compendium, 115

**E**
Eagle stones, 120
Egypt, 34, 38
"Empirical drugs", 22
Endogenous animal concretions, 115, 118
*Ens astral*, 4
Ettinghausen, R., 105, 107–110
*Etymologiarum*, 126
European apothecary shops, 130
European medical authors, 122
European Renaissance poisons rate, 56

**F**
Fifteenth-century toxicology, 162
*Five Entia* theory, 3–4

Florence, 79–81, 80*f*
Fluorspar, 85
Fossil sharks' teeth, 125–128, 127*f*
    Lapis de Goa, 131–132, 132*f*
    provenance of teeth, 129–130, 131*f*
    tableware, 128–129

**G**
Galen, 103
Galenic medicine, 95
Galenic Purgatives, Van Helmont criticism of,
    95–98
Georgius Agricola, 83, 84*f*
    *Bergsucht* and causes, 87–88
    conflict of interest, 90
    *De Re Metallica*, 85–86
    diseases in miners and prevention, 86–87
    education and early life, 83–85
    third wave of mining in 20th century, 88–90
Gessner, Conrad, 130
Glorious age of bezoars, 117
Glossopetra pendant, 128
Glossopetrae. *See* Fossil sharks' teeth
Goa Stone. *See* Lapis de Goa
Gold-plated ball, 131–132
Grand Duke Ferdinando I, 79–81
Grass snakes (*Natrix natrix*), 152
"Greater circulated salt" solvent, 98–99
Greek medical works, 121
Greek of Pseudo-Dioscorides, 48
Greek scientific and medical treatises, 45–46
Greek text, 47–48
Greek treatises, 48
"Griffin's claws", 128
Guaineri, Antonio, 162
Gypsum, 71–73

**H**
Hakim Seyed Bin Hasan Bin Seyed Mahdi,
    28
Hayāt al-Hayawān, 35
Hebrew translations and circulation, 39–40
*Helleborus niger.. See* Black hellebore
    (*Helleborus niger*)
Hellenistic and Roman apocryphal lapidaries,
    118
Hellenistic time, 119
Hemlock (*Conium maculatum*), 27–28
Henri de Mondeville (French surgeon), 152,
    154–155
Henry II of France, 71
Herbal poisons, 55–56
Hermetic literature, 118

Printed in the United States
By Bookmasters